COMPOSITE POLYMERIC MATERIALS

COMPOSITE POLYMERIC MATERIALS

R. P. SHELDON

Senior Lecturer in Polymer Science,
School of Polymer Science, University of Bradford,
Bradford, Yorkshire, UK

APPLIED SCIENCE PUBLISHERS
LONDON and NEW YORK

APPLIED SCIENCE PUBLISHERS LTD
Ripple Road, Barking, Essex, England
Sole Distributor in the USA and Canada
ELSEVIER SCIENCE PUBLISHING CO., INC.
52 Vanderbilt Avenue, New York, NY 10017, USA

British Library Cataloguing in Publication Data

Sheldon, R. P.
  Composite polymeric materials
  1. Polymers and polymerization
  2. Composite materials
  I. Title
  620.1′92   QD381.8

  ISBN 0–85334–129–X

WITH 26 TABLES AND 59 ILLUSTRATIONS

Photoset in Malta by Interprint Limited.
Printed in Great Britain by Galliard (Printers) Ltd. Great Yarmouth

To my family

ACKNOWLEDGEMENTS

I would like to record my appreciation of the assistance given by colleagues and students of the School of Polymer Science, University of Bradford, and the staff of the Department of Educational Technology at the same institution for preparing photographs. In addition, thanks are due to Mr J. Golding (Ciba-Geigy) for making available data on epoxy resin systems and to Dr D. C. Allport (ICI Ltd) for information relating to reinforced reaction injection moulding.

PREFACE

With the growing demand for lightweight, high performance materials, coupled with the escalating cost of energy and petrochemical feedstocks, the last few years have seen a tremendous surge of interest in composite polymers. These are not only displacing traditional engineering materials in many of their applications but are also creating new applications unique to themselves. So rapid have been the changes in both the growth and understanding of the new materials that many of the books published to date have, by necessity, either been on some narrow division or specialised facet, or have been a compilation of individual papers read at one of the many international conferences which have been held on the subject. The present book is intended to provide a broad treatment of the whole field of polymer composite materials in such a way that it should be of interest to practising polymer and materials scientists, technologists and engineers, and to advanced course and research students. In addition, it should be of value to final-year degree students taking special options in applied chemistry, colour chemistry, textiles, physics, etc.

Following an introduction to the nature of composite polymers, the book goes on to provide a more detailed account of polymers and fillers commonly used to form composites, their mechanical, physical and chemical properties, information on the rheology, compounding and processing of the various systems, and to outline the nature of polymer blends and copolymers. The final chapter is devoted to a range of selected aspects of the subject including examples of important composite systems, reinforced and syntactic foams, hybrid fibre reinforcements, polymer cements and, finally, an assessment of possible future trends in composite polymeric materials in the light of current developments.

R. P. SHELDON

CONTENTS

ix

GLOSSARY AND ABBREVIATIONS

ABS	Acrylonitrile–butadiene–styrene terpolymer
A-glass	Type of glass (other glasses include B,C,D,E,S, etc.)
Alloy	Polymer blend
Aramid	Aromatic polyamide
Aspect ratio	Length/diameter of fibre
ASTM	American Society for Testing of Materials
A-stage	First stage in thermoset cure
BET	Brunauer–Emmett–Teller (surface area technique)
BMC	Bulk moulding compound
BPF	British Plastics Federation
BS	British Standard
B-stage	Second stage in resin cure
CA	Cellulose acetate
CAB	Cellulose acetate-butyrate
CF	Carbon fibre, continuous filament
CFRP	Carbon fibre reinforced plastic
CNR	Carboxy nitroso rubber
CR	Neoprene (chlorinated rubber)
CSM	Chopped strand mat
C-stage	Third stage in resin cure
CV rubber	Viscosity-stabilised rubber
DAP	Diallyl phthalate
DBP	Dibutyl phthalate
DMC	Dough moulding compound
DSC	Differential scanning calorimetry
DTA	Differential thermal analysis

EPDM	Ethylene–propylene tercopolymer
EPM	Ethylene–propylene copolymer
EPR	Ethylene–propylene rubber
ESC	Environmental stress cracking
ESD	Equivalent spherical diameter
EVA	Ethylene–vinyl acetate copolymer
Fish eyes	Globular structural impurities in PVC, etc.
FRP	Fibre reinforced plastic
FW	Filament wound
Gel time	Time to gelation in curing reaction
GF	Glass fibre
GRP	Glass reinforced plastic
GRTP	Glass reinforced thermoplastic
HDPE	High density polyethylene
HDT	Heat distortion temperature
HET acid	Hexachloro-endo-methylene tetrahydrophthalic acid
HIPS	High impact polystyrene
Hybrid	Mixed filler (especially fibre) system
IEN	Interpenetrating elastomeric network
ILSS	Interlaminar shear strength
IPN	Interpenetrating polymer network
ISO	International Standards Organization
IUPAC	International Union of Pure and Applied Chemistry
K-value	Heat insulation (and other) index
LCST	Lower critical solution temperature
L/D	Length/diameter ratio (aspect ratio, extruder screw)
LDPE	Low density polyethylene
M,(MW)	Molecular weight
MBS	Methacrylate–butadiene–styrene terpolymer
M/F	Melamine/formaldehyde resin
MFI	Melt flow index
MVT	Moisture vapour transmission
MWD	Molecular weight distribution
NBR	Nitrile butadiene rubber
NDT	Non-destructive testing
NOL	Naval Ordnance Laboratory
NR	Natural rubber
PAN	Polyacrylonitrile
PC	Polycarbonate
PCC	Polymer concrete cement

PCTFE	Polychlortrifluorethylene
PET	Polyethylene terephthalate
P.F.	Power factor
P/F	Phenol/formaldehyde resin
phr	Parts per hundred of resin
PIB	Polyisobutylene
PIC	Polymer impregnated cement
PMMA	Polymethyl methacrylate
PP	Polypropylene
PPO	Polyphenylene oxide
PPS	Polyphenylene sulphide
PRI	Plastics and Rubber Institute
PS	Polystyrene
PTFE	Polytetrafluorethylene
PTMT	Polytetramethylene terephthalate
PV	Pressure (or load)–velocity product (wear)
PVC	Polyvinyl chloride
RAPRA	Rubber and Plastics Research Association
RF heating	Dielectric heating
R.H.	Relative humidity
RIM	Reaction injection moulding
RP	Reinforced plastic
RRIM	Reinforced reaction injection moulding
RTP	Reinforced thermoplastic
RTV	Room temperature vulcanising (rubber)
SAN	Styrene–acrylonitrile copolymer
SBS	Styrene–butadiene–styrene block copolymer
SIN	Simultaneous interpenetrating network
SIS	Styrene–isoprene–styrene block copolymer
S–N	Stress–number of cycles to fatigue (curve)
S.P.	Softening point
SPE	Society of Plastics Engineers
SPI	Society of Plastics Industry
SMC	Sheet moulding compound
Staple	Short length fibre
T_g	Glass transition temperature
T_m	Melting point
THF	Tetrahydrofuran
TPE	Thermoplastic elastomer
TPR	Thermoplastic rubber

TPU	Thermoplastic polyurethane
UCST	Upper critical solution temperature
UD	Unidirectional (aligned fibres)
U/F	Urea/formaldehyde resin
UFC	Universal fatigue curve
UTS	Ultimate tensile strength
WFS	Wet flexural strength
WLF	Williams–Landel–Ferry (equation)
WR	Woven roving
Z-blade	Z-shaped blade (also Σ blade)

Chapter 1

NATURE OF POLYMER COMPOSITES

INTRODUCTION

The need for new materials has led in part to today's drive for more and better polymer composites. Light-weight high-performance engineering plastics are ousting metals in many applications, just as synthetic polymers as a whole have already displaced, in total volume production, steel, hitherto the major structural material. However, there are more compelling reasons why the new materials are gaining dominance. The continuing development and increasing sophistication of technology, and the claims for a higher standard of living by people in the so-called developing countries are now drawing on the world's resources at an alarming rate. Why all this should encourage a growth in the importance of polymer composites which themselves depend on a supply of basic resins and synthetic elastomers obtained from petrochemicals may not be clear.

There is simply a greater potential for growth through a greater production of the basic synthetic polymers themselves. At the present time over half of all the oil produced is used to provide energy for transportation and the bulk of the remainder is used for purposes other than the production of petrochemicals. More and more the question is being asked as to whether other energy sources should not be more actively developed so that material resources, at least in a convertible form, may be conserved. The increased use of petroleum oil and natural gas for the production of synthetic polymers and at least a corresponding decrease in their use as direct sources of energy would go a long way to meet these aspirations. Particularly would this be the case if, the stocks were further extended by incorporation of massive quantities of cheap and low-energy-related inorganic and mineral fillers of which there is no serious shortage. Of course, there is the question also of the irrecoverable dissipation of energy even in the production of polymers from the basic chemicals but there is still an advantage, especially on a volume rather than a weight basis, compared with the energies involved in the production of other materials, as shown in Table 1.1. Going further in this

TABLE 1.1
ENERGY REQUIREMENTS FOR THE PRODUCTION OF DIFFERENT MATERIALS

Material	Energy	
	MJ/kg	KJ/cm^3
Bottle glass	18	41
Low density polyethylene	69	64
High density polyethylene	70	67
Polypropylene	73	68
Polyvinyl chloride	53	69
Polystyrene	80	84
Polyurethane	130	100
Polypropylene/30% glass fibre	90	100
Polyester/30% glass fibre	90	150
Phenoplast	150	200
Steel	45	350
Aluminium	> 200	> 540
Brass	95	600

direction of energy saving, which in fact is giving rise to quite a lot of concern within the polymer industry at the present time, it might well be that we shall see a shift in emphasis in the standing of some of the presently important polymers, such that the more versatile polymers, capable of accommodating appreciable quantities of filler without deterioration in properties or even with improvement in properties, may come to the forefront. The same theme can be explored with respect to processing. Indeed, already, polymeric systems historically associated with one labour-intensive or energy-costly fabrication process are being moulded in the same or alternative forms by newer and intrinsically less expensive rapid-production techniques, thereby gaining a competitive edge on other previously preferred materials. Compared with non-polymeric materials, polymers as a class have the advantage of being moulded by rapid and efficient techniques capable of mass production of finished articles.

Environmental pressure groups are urging the reclamation of plastics, which can be done typically by the manufacture of polymer blends, another type of composite polymer, although at the present time the costs of large-scale recovery can be fairly high. However, it is being done and no doubt suitable legislation, financial encouragement by governments, or changes in the pattern of material evolution and prices, might well further extend this trend.

Now that the importance of polymers and their composites from the resource and applicational point of view has been outlined, it is relevant to consider them in terms of the tonnage involved. At the moment, something like 50 million tonnes of synthetic polymers are produced per annum, the figure being expected to double in the next 20 years. Bearing in mind the difficulty, as will be considered later, of deciding just what is a filler (i.e. whether one should include pigments with extenders and reinforcing agents) one can assume that fillers probably account for around 10 per cent of this figure, many plastics now containing nearer 40 per cent and the overall proportion increasing every year. In money terms it is estimated that in the USA alone the amount spent on fillers for polymers will reach 1·4 million dollars by 1985, a higher proportion being spent on fibre reinforcement, although in tonnage, inorganic and mineral fillers will still dominate the market. The actual use of filled polymers covers many areas but in order of production importance, they are to be found in transportation, marine, construction, corrosion-resistant, electrical and electronics, appliance and business equipment, consumer and recreation, aircraft and aerospace components, and in a number of miscellaneous applications. The growth rates in these various categories are not the same, so that the relative order may well change.

Summarising, we can now appreciate the present and growing importance of composite polymeric materials and in so doing accept the need for a closer look at their nature, properties, and more detailed applications. Before doing this, however, let us look at the history of their development in order to put into perspective the potential for the future, to be outlined in the final chapter.

HISTORY

The first man-made composites based upon polymers appeared in about 5000 B.C. in the Middle East where pitch was used as a binder for reeds in boat-building (Table 1.2). It is still being used for this purpose in this

TABLE 1.2
HISTORICAL DEVELOPMENT OF POLYMER COMPOSITES

Date	Material
ca. 5000 BC	Papyrus/pitch (boats)
ca. 1500 BC	Wood veneer
1909	Phenolic composite
1928	Urea formaldehyde composite
1938	Melamine formaldehyde composite
1942	Glass reinforced polyester
1946	Epoxy resin composites
1946	Glass reinforced nylon
1951	Glass reinforced polystyrene
1956	Phenolic–asbestos ablative composite
1964	Carbon fibre reinforced plastics
1965	Boron fibre reinforced plastics
1969	Carbon/glass fibre hybrid composites
1972	Aramid fibre reinforced plastics
1975	Aramid/graphite fibre hybrids

country, more specifically in Wales by the descendants of the Celts who themselves had a connection with the Middle East, as it has been for perhaps 2000 or more years in the building of coracles for fishing. (Incidentally, it is of interest to note that the same material, pitch, is presently being assessed as a precursor for one of the most important components of ultra-modern reinforced plastics, namely carbon fibre.) Laminated wood dating to about 1500 B.C. has been found at Thebes and similar laminates based upon shellac resin have been known and used in India for at least 3000 years. In the first century A.D., Pliny was describing the use of wood veneers and so underlining the continuing nature of composites rather than the introduction of distinctly different and new types of composites. Proteinaceous resins such as are derived from glue and egg have found application in Europe, Asia, and America, and no doubt elsewhere, for well over a thousand years and were still in demand in the days of the alchemists, and far-seeing intellectuals like Leonardo da Vinci were quite clear as to their potential. On the filler side, it is of interest to note that glass fibre was known to the Phoenicians and was used by them in bottle making, but it was not until 1713 that it again gained momentary prominence when exhibited as a scientific curiosity at the Paris Academy. Even then the scientific world soon lost interest again until it was once more 'rediscovered' in the 1930s, being quickly adopted for incorporation in polymers to form composites, so emphasising the point made earlier that developments in one area can assist growth in another, this time to the mutual advantage of both glass fibres and unsaturated polyester resins.

The foundation of the modern polymer industry really began with the emergence of synthetic organic chemistry in the last century. In 1847, the Swedish chemist Berzelius prepared what was probably the first polyester resin, although the first commercial plastic (as distinct from the improvements which were being made in natural rubber and its technology to put it on the road as an engineering material) was not forthcoming until 1862 when Parkes introduced cellulose nitrate plastic. It is worth remarking that even at this early date, the polymer was strictly a compounded one, since it was blended with camphor. The next significant advance was made in about 1909 with the marketing of Bakelite resin, a composite of phenol-formaldehyde resin, discovered nine years earlier, with, typically, paper or cloth. This achievement draws attention to the point that, other than as a reinforced material phenol-formaldehyde resin, and indeed many other presently important polymers, would have been little more than a chemical curiosity. In the 1930s, Carleton Ellis, Carothers, Kienle and many others made substantial contributions to the synthesis of new polymers which have provided the backbone of the composites that have been most prominent in the last four decades. About the same time and shortly afterwards, there followed developments in many vinyl polymers, particularly the polyolefines, polyvinyl chloride, and polystyrene and then in polyurethane polymers, epoxide resins, etc., and more recently polycarbonate and acetal polymers with the later introduction, largely stimulated by the search for thermally stable polymers, of polysulphones, polypropylene oxide and sulphide, polyimide and related resins, aromatic polyamides and a whole host of exciting materials all striving for position which involves overcoming price disadvantages and establishing cost-effectiveness in terms of useful properties. Unfortunately, the price situation is not helped by the high costs of the most appropriate reinforcing fibres like carbon and boron which themselves are expensive to produce. However, as design demands grow both in breadth of application, and quantity within an application area, the above disadvantages are likely to be ameliorated.

So far, little has been said about the nature of the polymer composites other than indicating that they are mixed systems of polymers and fillers. The actual range of possibilities, so important for their use, is completely obscured by this trivial concept so it is worth while examining the classification of composites which at the same time illustrates geometrical morphology or arrangement of filler and polymer relative to each other.

CLASSIFICATION AND GEOMETRICAL MORPHOLOGY

It is probably true to say that all polymers contain some form of additive, ranging from small fractions of catalyst residue to large scale incorporation of, say, a mineral filler. The most important additives from our point of view are those introduced for some specific purpose and would therefore include (in volume order) fillers, plasticisers, colourants, reinforcing fibres, blowing agents, stabilisers against heat and sunlight, flame retardants, processing aids, and a final group of miscellaneous additives. Generally speaking, polymer composites, which may be regarded as multiphase materials of two or more components in which the polymer forms the continuous phase, can be considered as containing fillers or reinforcing agents, the function of the two frequently overlapping. Only where other types of additives are incorporated at such a level, typically over about 5 per cent, that they have a significant influence on, say, the mechanical properties, are they considered of being of comparable importance to conventional fillers. Thus for the purposes of this text, although we may at times regard some pigmented polymers or those heavily loaded with certain kinds of flame and smoke retardants as composites, most other polymer–additive systems would not be so included. Similarly, plasticised polymers in which one of the additives is a liquid of the conventional kind will be excluded from our consideration, even though in some cases a definite two-phase organisation might exist. Polymer cements in which the continuous phase is ceramic and in which the properties are essentially those of a modified ceramic rather than a modified polymer are again not strictly relevant to our purposes. However, with deference to the growing interest in all kinds of composites, including those based upon metals also, a brief mention will be made of polymer cements towards the end of the book (Chapter 7). On the other hand, some single component polymeric systems which exist in multiphase conditions, but not necessarily in crystalline and amorphous phases, exemplified by graft and block copolymers, will be included since their behaviour and the principles underlying this behaviour are similar to those of composites in general and polymer blends (which are also considered) in particular. In both cases the effect of addition of conventional filler will be examined.

Within these restrictions there are still many ways in which polymers and their constituent fillers can associate with each other in space. Probably the simplest organisation, conceptually, is that of the direct or contact laminate in which there are alternating layers of resin and sheet filler offering variability in structure and properties not only in the nature of filler and resin, but also in relative thickness and number with the further possibility that different layers may be of different materials, as, say, in a simulated-wood table top. To extend the scope the layers may have anisotropic properties in a mechanical sense, so that each layer may be arranged in a sequence in which the anisotropy is in the same, alternating or different directions, producing a final composite of quite different mechanical properties. Further, the layers may be of different porosity, varying from the continuous sheet capable of accepting little penetration of resin to an open weave which favours an intimate mixing of resin and sheet. The weave can vary considerably and for such a common fibrous reinforcement as glass may include plain, different twills, satin, leno, and mock-leno, or the filler may be a non-woven fabric, random orientation of loose fibres or, as in pultrusion for example, a layer of closely aligned parallel continuous fibres. All these geometries lend themselves to further complexity of organisation as can be easily imagined. To take just one instance, which is of technological importance, the fibre system may not just be of one type but can be a hybrid of different kinds of fibres. It should be recognised also that the sheet may be introduced as a narrow or ribbon version.

The geometry of the above examples is one of an essentially continuous form which can be cut or fabricated to size, but in addition polymer composites are frequently made up of discontinuous fillers being then loosely classified as particulate or fibrous. For purposes of characterisation, the three basic classes of the former are often considered as spherical, cubical, and tabloid but such a simple type of description masks the attendant complexities arising from some irregularity of shape, the presence of convoluted surfaces, and porosity. There is even the possibility of a change of shape during processing, if the filler is easily deformable either at room temperature or at the moulding temperature. Such could well be the case for rubbery fillers or low-melting-point solids. In the case of discontinuous fibres there are variables of fibre length and length distribution, surface geometry, and cross-sectional area, with the addition of other fibre-related geometries represented by wires and whiskers. Of the more important commercial fibrous fillers, whiskers with diameters of the order of 2×10^{-5} cm have the greatest surface area per unit volume, carbon fibre with its irregular profile being slightly finer than glass fibre, and boron fibre with a diameter about a hundred times that of a whisker having the lowest specific surface. Perhaps mention should be made at this stage of the lesser known bicomponent fibres, concentric or otherwise, made up of two different polymer types and themselves constituting a two-phase system. The opportunities for additional geometric variables through different spatial arrangements of fibrous fillers depend to some extent on their length but for short fibres and whiskers include random orientation in two as well as three dimensions. Longer fibres and continuous filaments allow such spatial arrangements as parallel wound fibres in planar orientation exemplified by the so-called cross-ply composites as indicated earlier. Another type of parallel location is to be found in pipes of circularly wound cylindrical reinforcements. Finally, as before, there are possibilities of mixed systems of different kinds of fibres and of fibres and particulate fillers. Examples of some of the above and other arrangements are shown in Fig. 1.1.

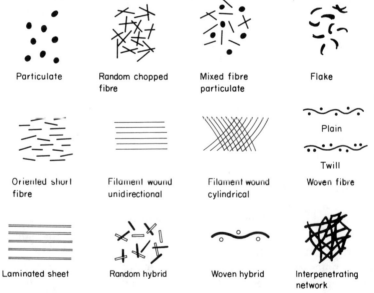

| Particulate | Random chopped fibre | Mixed fibre particulate | Flake |

| Oriented short fibre | Filament wound unidirectional | Filament wound cylindrical | Plain / Twill / Woven fibre |

| Laminated sheet | Random hybrid | Woven hybrid | Interpenetrating network |

FIG. 1.1. Examples of different composite geometrical arrangements.

In the composites just described we have implicitly been considering materials of two separate origins which have been physically produced by dispersing one phase (filler) in a continuous matrix (polymer). But there is a further class which extends this concept, being represented by what are called polymer blends or alloys. These are polymer composites in which, in addition to one phase being fluid at some time in the preparation, as with more or less conventional composites of the kind already outlined, the second phase may also be fluid, either as melt or as polymerising monomer. In this way, as will be seen later, a range of possible structures can arise including particulate ˙ dispersions, interlaminar distributions, bicomponent fibres as already mentioned, pseudo-fibres shaped by the fabrication process but not having the characteristic fibrous properties of conventional fibres, interpenetrating networks, etc. In addition they provide a facility which none of the other systems permit, in that there is now the opportunity for phase reversal, or inversion, depending primarily but not entirely on the relative concentrations of the two polymers, since relative viscosity at the fabrication temperature is also important. Thus from a state of a composite being continuous in phase with respect to one component there can be a change to a system which is continuous with respect to the second phase or to even one in which both phases are continuous. Needless to say, if the properties of the two polymers are different, then such phase changes can produce sudden changes in mechanical behaviour of the composite material.

In discussing discontinuous filler types of polymer composites it was said that they may be particulate or fibrous and that the fibres may be of different length, but no mention was made of the fact that particulate fillers also have a size distribution, which is narrow, for example, in manufactured glass beads, but wide in an as-received mineral filler. Since particle size has an important influence on properties either by disrupting flow patterns in fabrication or subsequent deformation processes, or, more importantly, through an effect on the interfacial contact area between resin and filler, some knowledge of particle size distribution is of importance in product design. This aspect has been of growing concern in recent years as technologists have attempted to incorporate increasing amounts of relatively cheap filler into polymers to extend their use or specifically improve properties. This might be towards obtaining more rigidity in the composite, to reduce overall thermal expansion or contraction, or to improve heat resistance by reducing the amount of the more thermally sensitive resin to a minimum. Generally speaking, the move towards improvements in these directions through increasing concentration leads to rheological difficulties in processing, but this is not always the case as will be reported presently. Complete analysis of size distributions experimentally, carried out by sieving or microscopy, and the use of data is not easy, involving as it does the application of mixing theory to undefined shapes.[1] For this reason it is usual to deal with spheres or to assume equivalent spherical diameters (e.s.d.s). Descriptions of particle size distributions can be made in different ways but typically may be as shown in Fig. 1.2. In these it must be

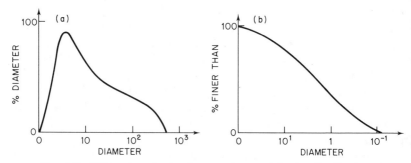

FIG. 1.2. Different methods of expressing particle size distribution.

remembered that the processing and purification of fillers from raw material often means a loss in the large or small size 'tails' so that the size distribution in the final moulded composite may be quite different to the size distribution of the original filler. In one example the effect of processing on breakdown of fillers was inferred by an enhanced nucleating effect on crystallisation by mica which has an easily cleaved crystal lattice.[2] A knowledge of size distribution or, better, the ability to tailor size-distribution will be appreciated when one realises that on the basis of choice of equal-sized spheres the theoretical maximum packing fractions are only 0·74 for hexagonal packing, 0·63 for random close packing, and 0·524 for simple cubic packing, thus implying considerable unused space. Because of this smaller spheres can, in principle, fit inside the resultant interstices, and even smaller ones between those of the second spheres, and so on. Indeed, it has been shown[3] that a mixture of spheres of diameter ratios 1:7:38:316 in the proportions 6:10:23:61 per cent by volume can reach a packing fraction of 0·95. What was also interesting in this work was that the mixture was still free flowing and that a polyester resin containing 50 per cent of the mixture was found to be just about as fluid as one of only 37·5 per cent using monodispersed spheres. The study of variable size distributions is still in its infancy but it is already being extended to include considerations of mixed fibre and particle size systems leading to such conclusions that in addition to economic advantages by massive incorporation of relatively cheap fillers into polymers, they can improve thermomechanical properties and that the presence of mobile particulate fillers like glass beads can improve the flow behaviour of otherwise fibre-reinforced only composites.

In the above, emphasis has been given to singly dispersed particles but in practice there is another aspect which is commonly encountered and one which has an important bearing on composite properties. This is the tendency for many particles to form agglomerates both before compounding and within the polymer matrix. All materials retain unsatisfied surface forces operating over distances of about 4–5 Å but falling off rapidly with increasing distance above this. They may be quite small in some cases as in aliphatic hydrocarbons, but in others especially in freshly-cleaved materials when there has not been time for contamination say by low-energy impurities or oxidation, they can be quite strong. Where they are strong, or where smaller particle sizes allow closer contact for overlap of force fields, appreciable aggregation of filler particles can occur. This is well illustrated by carbon black (Fig. 1.3) for

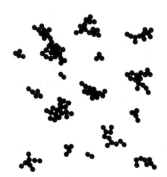

FIG. 1.3. Carbon particle dispersion (schematic, mag. $\sim \times 10^5$).

which there is considerable microscopic evidence of particle association. The tendency to agglomerate will of course be opposed by shear forces, and so the presence of agglomerates may depend upon processing conditions, and by the accidental or deliberate introduction on the particle surfaces of coating material with lower interaction forces, or with electrical charges which cause repulsion between particles. The latter is a feature of triboelectricity important in a number of polymer processing techniques. The occurrence of agglomerates is generally regarded as undesirable and therefore to be avoided wherever possible, giving rise as they do to optical imperfections, local heterogeneity of mechanical properties especially where locally trapped air pockets can lead to premature fracture of the composite, and to a lower packing fraction. Before concluding this section reference should be made to the intermediate, but somewhat related, state of dispersion where although possibly separately dispersed, the filler may exhibit high local (and conversely elsewhere, deficient) concentrations without being aggregated. This situation arises from some inefficiency in mixing or moulding, but it might be introduced by choice for producing some special optical effect.

To conclude, we have seen something of the intrinsic parameters associated with the use of fillers and more of this aspect will be mentioned later with regard to, for example, the influence of concentration on properties. There remains one parameter which has a profound effect on properties and that is the nature of the interface between polymer and filler, which will be examined now. From this we shall be able to consider ways of altering this interaction at will, and this will have important practical implications.

POLYMER–FILLER INTERFACE

Interfaces may arise in polymer systems for two reasons. On the one hand they may be the surfaces between two mutually insoluble and chemically distinct phases which are at equilibrium with respect to each other. On the other hand they may exist between two essentially miscible components which have not yet reached equilibrium, perhaps because of low diffusion characteristics of solid or high viscosity material. It is with the former equilibrium case that we shall be concerned here but it is important to remember that in a real mixing situation involving high viscosity polymers, the latter can also be of importance.

Just as in a chain where the weakest link dictates the strength properties of the whole, the mechanical behaviour of the composite reflects the interaction between the various phases. In short if there is no binding between the two, whether chemical or physical, as from secondary valence forces, the material response, for small strains at least, will be as if the matrix contained holes of shape identical to that of the filler. At higher strain the deforming matrix may impress itself on the more rigid filler, thereby producing a mechanical friction or related effect which will permit the filler to generate some growing influence on sample response.

If there is adhesion between the two phases then even at low strains stress transfer can take place across the interface, thus allowing the filler to share the stress and by so doing, as we shall see later, provide a reinforcing effect. If the interface is very rigid because of very strong interaction, a possibility arises of a deterioration in some properties which require some interface flexibility in order perhaps to dissipate excess energy. This can mean that for a particular application and expected force field, design of a useful composite may require some compromise in interfacial strength and interaction. One way of doing this, as will be seen, is to interpose a small proportion of a third component between the other two.

In the light of the considerable importance of interfacial interaction, it is not surprising to find that a great deal of effort has gone into practical and theoretical considerations of the interface, despite the intrinsic difficulty of experimentally studying what is essentially a two-dimensional domain rather than the three-dimensional bulk phase which is conventionally studied in science. It is not intended to summarise here the various techniques which are available for looking at surfaces, since they have been adequately dealt with elsewhere; suffice it to say that despite the tremendous advances which have been made in this area over the years, there is still a long way to go before any particular surface can be easily and comprehensively characterised.[4] As far as the present subject is concerned there are two aspects of relevance and although it is not appropriate here to go into any real detail, it is hoped that the following treatment will provide a working basis for appreciating the role of these two aspects and the factors affecting this role. The two subjects are wetting and adhesion on the one hand and polymer adsorption on the other. Considering first the former, it is necessary to invoke the concepts of surface and interfacial tension since these relate to the energies of surfaces.

The surface tension of a substance, typically a liquid, γ_{LV}, is a measure of the force which must be applied to the liquid to increase the surface area of the liquid and is defined in such a way that it is numerically equal to the work done in increasing the surface by unit area. Thus if a column of liquid is broken to produce two shorter columns each with a new end surface, the work done, per unit area (W_c), in overcoming the cohesion between layers of molecules will be:

$$W_c = 2\gamma_{LV} \qquad (1.1)$$

(For a substance in which the surface becomes a distorted structure of the bulk, as might apply to the case of a polymer, the relationship can become more complicated, a situation which we must presently ignore.) Correspondingly when a bond is formed between two dissimilar materials such as a liquid and a solid, the work of adhesion (W_A) will be superficially represented by the following equation enunciated by Dupré:[5]

$$W_A = \gamma_S + \gamma_{LV} - \gamma_{SL} \tag{1.2}$$

where γ_S and γ_{SL} are the surface tensions of the solid–air and solid–liquid interfaces respectively. In practice it is now considered that the solid is covered by a film of liquid on separating the bulk liquid and solid in breaking the interface, so it is usual to express the relationship in terms of a modified Dupré equation:

$$W_A = \gamma_{SV} + \gamma_{LV} - \gamma_{SL} \tag{1.3}$$

This equation may be converted into a more useful form by combining it with a relationship due to Young[6] which can be derived as follows. If we consider a drop of liquid at equilibrium on a solid surface as shown in Fig. 1.4 we see, ignoring an insignificant effect upon the surface arising

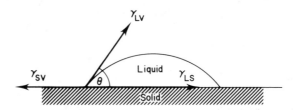

FIG. 1.4. Wetting of a solid (S) by a drop of liquid (L) in contact with its vapour (V).

from a vertical force component,[7] that the following holds:

$$\gamma_{SV} = \gamma_{SL} + \gamma_{LV} \cos \theta \tag{1.4}$$

Combining eqns (1.3) and (1.4) we arrive at the Young–Dupré equation:

$$W_A = \gamma_{LV} (1 + \cos \theta) \tag{1.5}$$

where θ is the contact angle, being a measure of the degree of wetting or interaction, taking a value of zero for a situation of ideal wetting. Assuming that interfacial breakdown will occur when the layer of liquid and solid are separated by a distance approximately equal to the distance over which intermolecular forces operate (see earlier) it is possible to get a figure for a theoretical interface bond strength. In practice, when this is done, bond strengths for filler–polymer composites, identifying the polymer as the liquid component, are orders of magnitude higher than experimentally determined values of adhesive bond strengths, even when so-called 'best ever' values are taken from repeat studies of adhesive joints (which are notorious in tensile testing procedures for being variable in value, highlighting further the point being made that bonding between polymer and substrate is a very variable quantity). In other words, there are factors other than apparent contact angle which influence adhesion or bonding between the two components. On face value, however, in order to get good bonding we need to minimise both contact angle and the other sources of weakness, so far unspecified. Regarding the latter, it is not difficult to speculate as to their nature but it is almost impossible to eliminate them. However, it is frequently possible by careful attention to detail to obtain guidance as to how they may be minimised, a point relevant to our considerations in trying to prepare useful composites. In the same way some guidance is available on ways in which contact angles may be reduced. Let us consider first of all the nature of the inherent weaknesses.

There are difficulties in making contact between two surfaces, which might typically be polymer and filler. These are derived from surface irregularities such as undulations, flaws, and microcracks many of which might be invisible to the naked eye. Even the smoothest surface, visually speaking, has been shown to contain depressions of the order of 10^2-10^3 Å, which means that below the resin surface there may well be indigenous stress concentrators. It might be imagined that under the influence of pressure during the preparation of an adhesive bond, say in a laminate, or during the moulding of a composite in general, the cracks would be filled with the flowing resin and so the influence of the stress concentrators reduced. However, reference to Poiseuille's equation for flow along a capillary[8] readily indicates that very long periods of time, many factors above normal moulding times, would be required for effective penetration of microcracks of the size indicated above. The equation in its simple form is given by:

$$V = \frac{\pi r^4 P}{8\eta l} \qquad (1.6)$$

where V is the volume rate of flow, r is the radius of the capillary, P is the hydrostatic pressure head, η is the viscosity of the liquid, high for polymer solutions and very high for melts, and l is the capillary length. For example, it has been calculated that for a capillary of diameter less than 10^{-5} cm a liquid of viscosity of 90 Pa.s under a moderate pressure of about 7×10^5 Pa would penetrate a distance comparable to the radius in five minutes, but with a polymer such as unvulcanized rubber, with a viscosity of about one hundred times that of the above, many hours would be required to produce the same penetration.[9] The presence of any air would increase this problem. In short, we see that a combination of high viscosity and small crack diameter more than cancel out high fabrication pressures which otherwise bring the two components together.

Another source of weakness leading to premature failure could arise from flaws within the matrix or filler which could produce breakdown not at the interface but within the bulk of either material. Impurities at the surface as well as affecting the wettability could give rise to occlusion of air or lead to vapour pockets, which in turn could again form stress concentration points. Indeed, such points appear to hold the key to the overall inherent weaknesses, arising, as they can, in a multitude of ways. They can form, for example, during moulding in which large changes of temperature might be involved, so that although at the moulding temperature no serious stresses are set up, the very act of cooling down to room temperature could easily produce tension within the material because of the different coefficients of expansion of the two components. Where the coefficients are not too dissimilar, as in some metal–epoxy systems, or where careful annealing or slow cooling has been carried out to allow matrix polymer molecules to adjust to some extent to the filler contraction, these stress concentration effects will be reduced. Compared with melt fabrication, curing systems for composite preparation in which, perhaps, a liquid monomer undergoes polymerisation, can produce even greater changes in volume, although fluidity of the polymer–monomer matrix will for a good part of the process permit continuous relaxation. The part use of fillers which counteract the shrinkage will also help. Probably most difficult to deal with, even for porous substrates, are the volume changes which occur during loss of liquid from polymer systems applied as solutions, this process combining as it does both shrinkage stresses and micro- and macro-void formation. The final reason why a bond between polymer and filler substrate might have a much lower strength than would be desired could be that incidental stresses are established non-uniformly under test or usage conditions. All these factors direct attention to suitable composite design, perhaps relating to morphological geometrical facets in relation to likely applicational stress patterns, as well as to careful fabrication methods.

Returning to the other complementary approach to good bonding, i.e. through reduction of contact angle, one can often achieve this by increasing the temperature, which can frequently increase the compatibility between molecules of the two media. However, an advantage gained this way might then be lost through stress concentrators set up by subsequent shrinkage on cooling as outlined above. The other potential advantages of increase of temperature lie in the possibility that this may remove volatile impurities, say for example, a film of water on the surface of the filler, or reduce matrix viscosity thereby helping penetration of surface irregularities by the polymer. Some indication of this behaviour can be seen in Fig. 1.5 which shows

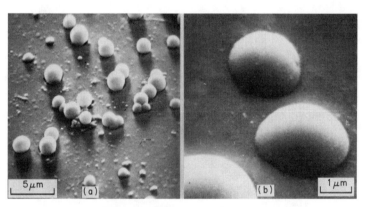

FIG. 1.5. Spreading of polyvinyl acetate on glass.[17] (a) Emulsion dried at room temperature. (b) Emulsion dried at 140°C (1 hour).

the effect of temperature on the spreading of polyvinyl acetate emulsion particles, sometimes used as a size in glass fibre production, on the surface of glass. In addition to changing temperature some roughening of filler surface can lead to a decrease in contact angle as was first shown by Wenzel,[10] and certainly this is a frequent practice in the preparation of normal adhesive joints, although to what extent the simultaneous cleaning of the surface by the abrasion also assists bonding is not clear. If, however, excessive roughening of the substrate surface is undertaken the advantage on bonding is lost through extensive irregularities, as already described. The final, and probably the most important way to produce a low contact angle, is to select the most suitable resin–filler system in the first place. It is here that we should turn to a separate approach to bonding which, although not foolproof, can provide a useful guide to wetting and the formation of good polymer–filler interaction, especially where polymer mixtures are concerned and where the surface of a filler may be contaminated by an organic impurity.

Following an extensive study of the wetting of a wide range of surfaces by different liquids, Zisman and his collaborators have been able to show that if one takes a homologous series of liquids, which will have slightly differing surface tensions, and determines the contact angle of each on a particular surface, a plot of $\cos \theta$ against surface tension produces a linear relationship.[11] Extrapolation of the line to the point where $\cos \theta$ becomes unity gives a so-called critical surface tension (γ_c) for the solid, as is shown in Fig. 1.6. If a non-homologous series of liquids is taken the

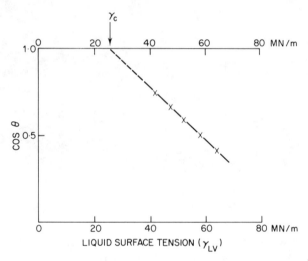

FIG. 1.6. Wettability and critical surface tension of a polymer surface using a homologous series of liquids.

extrapolation is found to be less precise, the value of critical surface tension now lying in a narrow band, the lower limit of which typically corresponds to the above value. Tables of critical surface tensions for various surfaces can now be drawn up as shown in Table 1.3. It is seen

TABLE 1.3
CRITICAL SURFACE TENSION OF POLYMERS

Polymer	Critical surface tension (γ_c) MN/m
Polyperfluoropropylene	16·2
Polytetrafluoroethylene	18·5
Polydimethylsiloxane	21
Polyvinyl fluoride	28
Polyethylene	31
Polystyrene	33
Polyvinyl alcohol	37
Polyvinyl chloride	39
Polymethyl methacrylate	39
Polyethylene terephthalate	43
Nylon 66	46

from this that low energy surfaces such as for hydrocarbons and fluorinated polymers exhibit low values of critical surface tension, whilst high energy surfaces such as for polyamides, have high values of critical surface tension. It is interesting to find that modification of a surface can alter the value of γ_c. For example, roughening of the surface of wood or treatment of the surface with certain liquids can raise the value of critical surface tension by a factor of about two. The significant consequence of these results for all polymers is that in so far as they are valid in a particular case, i.e. that they are characteristic of the surface and not of a film of foreign matter, once the critical surface tension is known then by implication one can presume that any liquid with a surface tension less than that of the solid's critical surface tension will wet the surface. This conclusion is borne out by the fact that most liquids will spread on the high energy surfaces of metals and inorganic substances, unless the surfaces are contaminated by low surface tension impurities which, of course, will easily come about for the very reason that they, too, will have surface tensions less than that of the substrate. A rather neat illustration of the situation is provided by the report that polyethylene can be used as an adhesive for an epoxy resin sheet which has a much higher critical surface tension, but that the corresponding epoxy resin is not suitable as an adhesive for polyethylene.[12] Thus now, in principle, we can match resin and substrate to achieve useful bonding, or alternatively can systematically modify one or both to achieve the same end. Some important ways in which a filler surface can be modified will be discussed in the next section but suffice it to say that two principal approaches have been made, the first being chemical modification of a low energy surface and the other, by the use of certain coupling agents. And again it must be remembered, that surface modification may take place unintentionally through the influence of environmental factors such as water vapour, oxygen, carbon dioxide, and the action of sunlight. Or again, small concentrations of impurities within the resin, otherwise having little influence on the properties of the resin, might diffuse and aggregate at the interface and in doing so compete with the resin for interaction with the filler surface. Examples of such impurities might be water, or possibly monomer residues and plasticiser. In all cases this behaviour could mean an inferior bond as first prepared or a deterioration which takes place over a period of time.

Drawing together the theoretical and experimental conclusions of surface studies in such a way that they may be of relevance to the fabrication of useful matrix–filler bonding in polymer composites, a list of recommendations can be constructed as shown in Table 1.4. Now that we have discussed the factors which are important to the wetting of surfaces by polymeric systems, it is of interest to have a look at the way that the polymer molecules are organised at the filler surface since this affects the relationship between bulk filler on the one hand and bulk polymer on the other, and in turn the way the composite as a whole responds to applied mechanical forces. However, it must be admitted

TABLE 1.4
FACTORS LEADING TO GOOD POLYMER-FILLER BONDING

 1. Low contact angle between polymer and filler
 2. γ_{LV} less than γ_C for spreading of polymer on filler surface
 3. Low viscosity of resin at time of application
 4. Increased pressure to assist flow
 5. High viscosity after application (cooling, curing, etc.)
 6. Clean and dust-free surface on filler
 7. Absence of cracks and pores on filler surface
 8. Moderate roughening of filler surface
 9. For impermeable fillers avoid solvent-based resins
10. Resin less rigid than filler (for reinforcement)
11. Coefficients of thermal expansion of components similar
12. Appropriate composite design (specific to intended application)

that unfortunately, not a great deal is known concerning the detailed arrangements of molecules at the interface and even much of this is largely inferred from data relating to studies of the adsorption of polymers from solution[13] rather than in the bulk. From this, it is presumed that the polymer forms a boundary layer on the surface of a thickness depending upon such parameters as chain length, chain flexibility, and the strength of intermolecular bonding which exists between atoms or groups of atoms in the polymer and on the surface, the latter, as already indicated, not necessarily being the same as within the bulk of the filler. Examples of possible thicknesses which have been reported are of 100 Å say for polyester on glass and 170 Å for a poly-urethane on carbon. In the process of adsorption, a dynamic situation is envisaged in which polymer segments are continuously exchanging substrate site points with each other, but leading to a progressive coverage of the surface with time, the actual behaviour depending also on the intensity of interaction. Since such a picture implies the need for a finite time for attainment of equilibrium, allowance must be made for this if the moulding conditions, perhaps involving high temperature, conversion of low-molecular-weight monomer to polymer or loss of solvent, are much different to applicational temperature conditions. To what extent the adsorbed polymer interacts with the free polymer is again not known, but is possibly a function of the strength of bonding, the flatter polymer configurational profiles tending to reduce molecular interaction with the non-adsorbed molecules. The strength of bonding will vary from that of weak van der Waals'/London type forces at one end of the scale to chemisorption forces, equivalent to primary chemical bonds, at the other. Examples of the latter, which will be less common, are believed to exist in systems such as carbon–natural rubber, silica– and glass–silicones, and clay–aminoplasts. In the first case, at least, they are so strong that on deforming a filled rubber, the corresponding contributory increase in the distances between carbon particles may in some cases be achieved not by customary chain unfolding or interchain slippage, but, because of the high strength, by actual fracture of primary carbon–carbon bonds in the polymer chain itself, thus leading to polymer chain degradation. This is reflected in a strain-softening effect and might be a factor of the so-called Mullins Effect in compounded elastomers (see later). Other evidence for strong interaction comes from thermal studies. If the change in glass-transition temperature (see Chapter 2), which is related to molecular chain mobility in a polymer, is measured for a filled polymer system it is often found that this has increased relative to the value for the unfilled material. When one realises that what is recorded is an average value for bound and unbound polymer, then one is led to the conclusion that the adsorbed polymer has undergone a transition temperature change of many degrees, perhaps as great as 100°C. It should nevertheless be mentioned that there are examples where the overall glass transition temperature of a polymer composite has decreased, but even this is not inconsistent with strong adsorption, if it involves preferential adsorption of the higher molecular weight component of the polymer which would have a greater contribution to the transition temperature of the un-filled polymer, or if the process of adsorption in some cases could pro-duce an increase in the free volume in the residual polymer. It is of interest to note, too, that the fall in chain mobility is reflected directly in the development of asymmetry in dynamic mechanical loss curves.[14]

Before leaving this section mention should be made of some of the other parameters which affect adsorption. However, as stated earlier, since most studies on adsorption have been made with solutions of polymers, the significance of the results with respect to the behaviour of bulk polymers is not clear, since in all cases the solvent molecules will be competing for recognition. Thus, although it has been found that increase in temperature can result in a decrease in adsorption, as might be imagined in a dynamic situation, there are also reports to the contrary. Similarly, conflicting indications have been found for the effect of change in polymer molecular weight.

COUPLING AGENTS

In order to establish an improved interaction between filler and polymer, it would appear from the previous section that attention should be paid to modifying either the polymer or the filler. Some ideas have been given as to how this might arise incidentally. Examples of ways in which it can be done deliberately will now be outlined. Where the substrate might be polyethylene, this has been done by means of strong oxidising agents which produce more polar groups on the surface; a similar effect is obtained for polytetrafluoroethylene by treatment with a sodium metal–ammonia system. For carbon and graphite fibres, oxidising agents such as sodium hypochlorite or dilute nitric acid produce active phenolic and carboxylic acid groups, although it might well be that some etching or cleaning of the surface also occurs, since this too, would lead to the observed increased interaction with a resin matrix. Metals may be activated by treatment with phosphoric or nitric acid and here, again, there is an etching effect. Etching is not confined to non-polymeric fillers since polyesters, which are known to be susceptible to amine attack in this way, may be similarly treated to improve bonding.

Despite a number of examples of this kind, the most important way of modifying filler surfaces is by the use of certain additives known as coupling agents, and this method is applicable to most polymer composites.[15] Even at very low concentrations (e.g. ~ 0.1 per cent), silane coupling agents, which are the best-known coupling agents, give rise to significant improvements in mechanical properties. The role of the additive is not altogether clear in that although there is a general correlation between efficacy and chemical structure through expected interaction or reaction with the molecules of the matrix, there is the possibility that they may also act by preventing ingress, say, of moisture, or making up for what might otherwise not be a perfectly clean filler surface. Another role may be in providing a means for stress relaxation in situations of different dimensional changes of matrix and filler when the composite is deformed. A separate use of additives which in general nature appear to be identical or at least similar to coupling agents and will be considered later, is in improving dispersion and rheological properties of polymer matrixes, as protective sizes and to prevent an otherwise restraining influence of filler in cases of in-situ polymerisation.

The general relationship between silane structure and applicability is apparent in Table 1.5 in which it is seen that there is more than a

TABLE 1.5
SOME COMMERCIAL SILANE COUPLING AGENTS

Coupling agent	Formula	Polymer system
Vinyltrichlorosilane	$CH_2{=}CH{-}SiCl_3$	polyester
γ-aminopropyltriethoxy-silane	$NH_2{-}(CH_2)_3{-}Si(OC_2H_5)_3$	epoxy resin
Allyldichlorosilane-resorcinol	$CH_2{=}CH{-}CH_2{-}SiCl_2O$ ⬡—OH	general

suggestion of specific chemical action between the various terminal groups on the silane and groups on the polymer, whilst the opposite terminal hydroxide will have strong interaction with the filler. The situation for glass is shown in Fig. 1.7. In one case, that of amino-terminated silane, it is probable that there is also a catalytic contribution

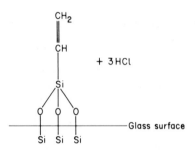

FIG. 1.7. Schematic coupling action of a silane (vinyltrichlorosilane) to glass.

to curing of the epoxide. Particular use is made of silanes in applications involving glass fibre and beads, mica, and calcined clays. Closely following the silanes in importance, and indeed at present enjoying a more rapid increase in use, are the organic titanates, exemplified by isopropyl tri(dioctylpyrophosphate) titanate which binds to an inorganic surface by hydrolysis.[16] These materials are finding application to a wide range of particulate and fibrous fillers. Other coupling agents include chromium complexes (which have a long history of use with glass), phosphorus-containing compounds, long chain aliphatic carboxylic acids, and certain amines.

A different type of coupling agent is provided for fairly specific systems by certain polymers, usually copolymers. In their application as block and graft copolymers to link what are otherwise immiscible homopolymers, usually of types identified with the nature of the blocks in the copolymers, their role appears to be fairly clear in that they provide a basis for mutual compatibility for the homopolymers both in the solid state and in solution. Polymers are also finding a role of offering a coupling action between two other polymers having no immediate identity with the coupling agent. An example of this is in the treatment of aramid fibres with polyurethane when used as reinforcing agent in epoxy resin systems. This leads to a separate use of polymers as coupling agents. Although many conventional agents improve the stiffness and tensile strength properties of composites, they frequently have no useful influence on shear and impact behaviour, some additives actually leading to a deterioration in these latter properties, whilst at the same time giving rise to strong bonding. Early recognition of this led to the suggestion that a certain amount of debonding propensity might be advantageous in some cases. However, there is a growing body of opinion that an even better way to obtain this flexibility is to use polymer-coated fillers, prepared for example by treatment with a polymer solution which is then dried. Some examples of such systems are silicone rubber-coated carbon fibre, epoxy-coated boron, and polyester-coated glass for polyester reinforcement. The use of polyethylene-impregnated particulate fillers has been known for many years. In these systems the actual thickness of the coating is important with useful thicknesses being of the order of microns, and therefore thicker than for the strength- and stiffness-promoting silane coupling agents, which although variable might typically be no greater than, say, 0·05 micron on a fibre of diameter 10 microns, i.e. corresponding to a level of addition of about 1 per cent.

Finally, mention is made of some of the other coating agents which are applied to fillers for reasons other than direct improvement of mechanical properties. One is for improving flow and dispersibility. There are two aspects to this action, the first being in regard to easing flow in moulding operations, and the other to permit higher loadings of a filler in a composite. The latter produces not only a greater economic advantage for composites but also allows for an enhancement of applicational properties, as exemplified by the use of silanes in assisting heavier loadings of the flame-retardant filler, aluminium trihydrate, or to provide higher concentrations of glass fibre in the important RRIM process to be described later. Incidentally, the ability of a filler to be dispersed in a resin system is characterised by what is known as its 'oil absorption' coefficient, being a measure of the amount of linseed oil needed to convert a dry filler into a stiff paste. Calcium carbonate has one of the lowest coefficients for conventional fillers, but even this, as for other fillers, can be modified by use of appropriate coating agents.

REFERENCES

1. FERRIGNO, T. H., in Katz, H. S. and Milewski, J. V. (Eds.), *Handbook of Fillers and Reinforcements for Plastics*, Chapter 2, Van Nostrand Reinhold, New York (1978).

2. IBBOTSON, C., PhD Thesis, University of Bradford (1976).
3. MCGREARY, R. K., *J. Am. Ceramic Soc.*, **44**, 513 (1961).
4. ERICKSON, P. W. and PLUEDDEMANN, E. P. in *Interfaces in Polymer Matrix Composites*, Plueddemann, E. P. (Ed), Academic Press, New York (1974).
5. DUPRÉ, A., *Théorie Mécanique de la Chaleur*, Paris (1869).
6. YOUNG, T., *Phil. Trans. Roy. Soc.*, **84** (1805).
7. CHERRY, B. W., *Polymer Surfaces*, Cambridge U.P., Cambridge (1981).
8. POISEUILLE, J. L. M., *Comptes Rendus*, **11** 961, 1041, **12**, 112 (1840).
9. WAKE, W. C., in Eley, D. D. (Ed), *The Rheology of Adhesives*, p. 193, Oxford U.P., Oxford (1961).
10. WENZEL, R. N., *Ind. Eng. Chem.*, **28**, 988 (1936).
11. FOX, H. W. and ZISMAN, W. A., *J. Colloid Sci.*, **7**, 109 (1952).
12. SHARPE, L. H. and SCHONHORN, H., *Adv. in Chem.* Series, No. 43, A.C.S. 189 (1964).
13. See e.g. LIPATOV, Y. S., *Physical Chemistry of Filled Polymers*, Transl. Int. Polymer Sci. and Tech. Monogr. No 2, Rubber and Plastics Res. Assoc., Shawbury (1979).
14. ARAS, L., SHELDON, R. P. and LAI, H. M., *J. Polymer Sci: Polymer Letters*, **16**, 27 (1978).
15. PLUEDDEMANN, E. P. (Ed), *Interfaces in Polymer Matrix Composites*, Academic Press, New York (1974).
16. MONTE, S. J. and SUGARMAN, G., in Seymour, R. B. (Ed), *Additives for Plastics*, Vol 1, Academic Press, New York (1978).
17. HODGSON, I. and SHELDON, R. P., Unpublished results.

Chapter 2

POLYMERS AND FILLERS

NATURE AND SYNTHESIS OF POLYMERS

In order to appreciate in which way and to what extent the properties of polymer composites differ from those of the parent polymers, it is necessary to understand something of the fundamental nature of polymers. In addition, since their nature may depend upon the way the polymer was made in the first place, different methods in principle being able to produce materials of essentially the same chemical nature but having different physical properties, it is important to consider the details of the actual synthesis of polymers.

Polymers are substances composed of long-chain molecules, being, in the context of the present interest, primarily organic, that is, they are based predominantly upon carbon chains. They exist in Nature in the form of, for example, cellulose, proteins, and isoprene polymers. The natural polymers are not usually in a pure form but are associated with other matter; indeed, wood itself is a composite polymer of mainly cellulose and lignin. In Nature, the polymers are formed from comparatively simple low molecular weight units by the action of appropriate and efficient catalysts known as enzymes. Similarly, synthetic polymers are prepared, much less efficiently, by means of a wide range of initiating and catalytic systems, also operating on, simple monomeric starting units. The monomers are obtained mainly as petrochemicals at the present time, but alternative sources are available from raw materials such as coal, sugar, wood alcohol, and furfural. Some of these are already growing in importance because of restrictions on the supply and other uncertainties in the future of petroleum and natural gas.

The polymers are prepared by two principal processes, step-growth polymerisation and addition polymerisation, the latter being sub-divided into free radical and ionic mechanisms. The two major classes of polymers formed by either process are the thermoplastics, which undergo softening or melting on heating and therefore are amenable to liquid-flow shaping techniques, and thermosetting polymers in which primary bond chains are formed between the main chains of the polymer, thus putting restrictions on any possible thermal-forming operations following initial cure. The continuous availability of a melt pool has meant that for many purposes the former class of polymer has been preferred for long moulding runs, but over recent years a growing sophistication in techniques has led to a minimisation of the disadvantages of the thermosetting class of polymers. In step-growth preparation, typified by the reaction between a difunctional diol and a difunctional di-acid, or a hydroxy acid with itself, usually with an acid catalyst, exemplified by the polycondensation of nylon, the average molecular weight of the system increases progressively with time. The reaction is often accompanied by the elimination of a simple molecule such as water, but in some cases, involving, say, a ring-opening reaction or the formation of a poly-urethane, no incidental elimination product is formed. High average molecular weight, required for most applications where mechanical properties are important, is achieved by long reaction times, precise stoichiometry, and by the use of interfacial polycondensation approaches. Control of molecular weight to prevent too high values for convenient processing can further be obtained by addition of monofunctional reactants in the correct proportion. Examples of important polymers prepared in this way are presented in Table 2.1.

Addition polymerisation is carried out with monomers which contain unsaturated groups, the actual process being determined to some extent by the nature of the monomer. In free-radical polymerisation a free-radical generator, such as benzoyl peroxide, is used, this undergoing homolytic fission by, for example, heating. The radical combines with a monomer molecule to provide an active site for subsequent polymerisation. Anionic polymerisation is applicable to monomers which have a strong electronegativity and which can form carbanions under the influence of catalysts such as butyl lithium or an alkali metal. Correspondingly, cationic polymerisation takes place through carbonium ions obtained by use of Lewis-type acids, e.g. boron trifluoride. As indicated, some polymerisations are specific to a particular monomer, whereas other monomers are less selective. Thus whilst vinyl ethers

TABLE 2.1
SOME IMPORTANT SYNTHETIC POLYMERS

| Thermoplastics | | Thermosets | |
Step-growth	Addition	Step-growth	Addition
Polyamides	Polyethylene	Phenolic resins	Styrene/polyester
Polyimides	Polypropylene	Melamine resins	Synthetic rubbers
Acetal resins	Poly(4-methyl pentene–1)	Urea resins	Allyl resins
Polycarbonate	Polyvinyl chloride	Polyurethane	
Polyethylene terephthalate	Polyvinylidene chloride	Silicones	
Polyurethane	Polyvinyl acetate	Alkyd resin	
Silicones	Polymethyl methacrylate		
	Polystyrene		
	Polyphenylene oxide		
	Polysulfones		
	Polytetrafluoroethylene		

require cationic stimulus and vinyl esters need free radicals for reaction, polystyrene can be prepared from its monomer by all three methods. As in the case of step-growth polymerisation, multifunctionality can give rise to chain branching and interchain cross linking. It is interesting to note that one polymeric system which is very important in the context of polymers commonly used as composites, the polyester system, is actually formed in two steps, the first being a step-growth condensation involving an unsaturated di-acid and diol, and the second step being a free-radical addition reaction of the unsaturated groups with unsaturated styrene monomer later added as a solvent for the first-stage product. The second stage is a special example of solution polymerisation, though usually the solvent is non-reactive in this technique, which draws attention to a further variable in polymerisation reactions, that is, the nature of the reaction media. Other forms of addition polymerisation are to be found in bulk, suspension, and emulsion systems, each having significant implications as far as properties and later compounding procedures are concerned with reference to the production of polymer composites.

Before turning to a consideration of specific examples, it might be helpful to review some of the essential properties of polymers as this will serve as a basis for appreciating the modifying role of added filler. Average molecular weight and the profile of the molecular weight distribution, coupled with chain branching and cross linking, will all contribute, quite separately to the chemical nature of the polymer, to the way that it will respond to mechanical and physical stressing. There are other intrinsic molecular properties which will also govern behaviour. Those just noted are especially important for the fluid or molecular mobile state, since to a first approximation viscosity increases proportionally to a 3·5 index with respect to average molecular weight, for example, but other parameters can affect solid state behaviour. One of these is the inherent chain flexibility. All substances exhibit molecular movement at temperatures above 0 K but it becomes particularly noticeable as translational movement at the melting point for crystalline materials, or at the glass-transition temperature for amorphous materials, typically resulting in a thousand-fold drop in stiffness with amorphous polymers, and some intermediate factor for partially crystalline polymers at temperatures between the glass-transition temperature and the melting point. The more flexible the polymer chain, everything else being equal, the lower the temperature at which this occurs, i.e. the lower the glass-transition temperature of the polymer. Another important parameter is stereoregularity of the macromolecular chain, since this is the property which provides a basis for crystallisation of polymers. Since stereoregularity is closely bound to polymerisation mechanism, then both the nature of the reacting species and the particular method of polymerisation can influence the ability of the polymer to crystallise, and indirectly to retain some shape rigidity with increasing temperature. Since crystallisation is a nucleated process it is interesting to note at this point that the presence of the filler in a polymer composite, if the polymer is crystallisable, can in principle affect the structure of the polymer itself, in a physical sense, and hence further influence properties.

This physical organisation of the polymer, or morphology, as it is more usually termed, is sensitive not only to intrinsic factors of the kind just outlined, but also to external factors. Since the mechanical properties, in particular, are strongly influenced by morphology, it is appropriate that we examine the character of these latter factors. Perhaps the most important of these are mechanical stressing of the polymer, pressure, temperature, and treatment with certain liquids and solvents. The application of a mechanical force to a polymer can result in molecular and crystalline orientation as the two entities respond to the force field. The result might be either uniaxial orientation as in the production of fibres, or biaxial, as in the production of some films. In both cases orientation produces anisotropy of physical and mechanical properties. The effect of pressure is to encourage the molecules to accommodate the applied stress, again by rearrangement. This has a well-known effect of producing extended crystals in partially crystalline polymers. The most serious influence of applied stress and pressure, components of many fabrication procedures, is that under non-equilibrium conditions of rapid change in both, and in temperature, local stress inclusions can develop. These lead, as was mentioned in Chapter 1, to sites for premature failure. Correct thermal annealing can reduce or eradicate these and at the same time remove the sensitivity to certain liquids, which manifests itself as a phenomenon technically known as stress-corrosion. In connection with the effect of certain liquids, it is of interest for the special cases of polymer blends and copolymers, which are discussed later (Chapter 6), to draw attention to a curious influence on morphology, perhaps unique to such systems. If a two-component system of this kind is cast, say as a film, from a solution, the precise morphology can be a specific function of the casting solvent. For example, if the solvent has a greater compatibility with one component than the other, then there is a tendency for that component to form the continuous phase in the composite, whilst if a solvent is used which has a greater compatibility with the second component then a phase-inverted composite will be produced, (Fig. 2.1). Thus, in principle, merely altering the nature of the solvent, itself having no further influence on properties, can produce from the same material products of quite different mechanical and physical properties.

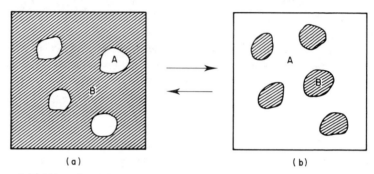

(a)   (b)

FIG. 2.1. Phase inversion for solvent-cast block copolymer or polyblend. (a) Good solvent for B; (b) good solvent for A.

Having briefly reviewed the basic nature of polymers in respect to structural effects on properties, let us now turn to a consideration of some of the polymers which have found a particular importance for use in composites.

THERMOSETTING AND THERMOPLASTIC POLYMERS[1]

The division of polymers into thermosets and thermoplastics has already been exemplified in Table 2.1. Historically, the thermosets, particularly in the form of phenolic resins, preceded the thermoplastics in terms of large-scale industrial production. However, although they constitute a vital part of the polymer picture, as far as tonnage production is concerned they are dwarfed by a factor of four to five times at present by the thermoplastic polymers. Informed opinion has it that, if anything, the ratio is more likely to increase than to decrease in the foreseeable future. In both classes a relatively few kinds account for the bulk of overall production. It is to these, and to a few more that are expected to play an increasingly important role in the immediate future, that the next section will be devoted. Many of the thermosets in particular, for obvious reasons, are compounded with filler at the initial stages of polymerisation, certainly before curing, but for the moment, as for the thermoplastics, emphasis will be placed on the reaction itself, the influence of filler in relation to polymerisation being considered elsewhere.

Thermosetting Polymers

Historically, the most important class of polymer composite is that of the phenolic resins or *phenoplasts* introduced on a significant scale in about 1912, i.e. about three years after the first marketing venture. These are the condensation products of phenols and aldehydes, the best known of the former being phenol itself and to a lesser extent one of a number of cresols, whilst the most important aldehyde is formaldehyde, followed by furfural. For practical reasons of industrial moulding, low molecular weight reaction products which are capable of being compounded with fillers or laminating sheets before final cure, are first produced. The two intermediates are known as novalaks or resols. In the case of the novalaks, a molar excess of phenol is reacted with formaldehyde under acid conditions to produce relatively low molecular weight linear products containing about five benzene rings and no reactive methylol groups. To obtain a resol an excess of aldehyde is reacted with phenol under alkaline conditions. In this, the number of benzene rings varies from about two (liquids) to about four (solids) and the product contains methylol groups. Technically, both are prepared in resin kettles, the time of reaction being approximately 2 to 4 h for novalaks and 1 h for resols. The novalaks are more frequently used in the form of moulding powders, at which stage they are compounded with appropriate fillers, whilst resols appear to be the preferred materials for laminating applications. Cross-linking in the case of novalaks is typically carried out by means of hexamethylene tetramine or paraformaldehyde. Both acids and bases will produce curing of resols, caustic soda being a not uncommon example of such a curing agent.

A later range of resins which compete with phenoplasts for many applications, having many superior properties but unfortunately being more expensive, was that of the *aminoplasts*, the reaction products of aldehydes and amines or amides. The two most important at the present time are the urea–formaldehyde and the melamine–formaldehyde resins. Like the phenoplasts, they are moulded under pressure. Resins of a low state of condensation based on a 50 per cent molar excess of formaldehyde for urea–formaldehyde resins and even more for melamine resins are prepared under slightly alkaline conditions in a resin kettle to produce moulding powders, laminating resins, and for other applications, adhesives and surface coatings. Although the initial products can be cured by heating in the temperature range of about 130–170°C, it is usual to employ mildly acidic catalysts to accelerate cross-linking.

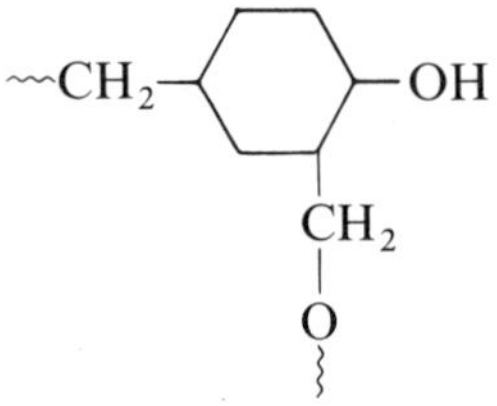

phenolic resin

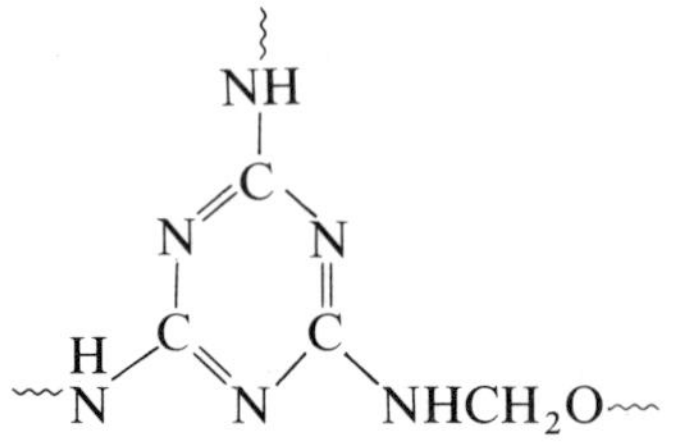

melamine formaldehyde resin

$$\sim CH_2\,NCO\cdot ONHCH_2NH\sim$$

urea formaldehyde resin

Perhaps the most versatile class of thermosetting polymers is that of the unsaturated *polyesters*. These are prepared by heating a diol, typically 1, 2 propylene glycol, with a mixture of di-acids or their equivalents in a resin kettle in the temperature range 150–200°C under an inert atmosphere of carbon dioxide or nitrogen. The acids are made up of an unsaturated component such as maleic anhydride or its *trans* isomer, fumaric acid, which provides the sites for further reaction, and a saturated acid, typically phthalic anhydride. Instead of the latter, an aliphatic acid such as adipic acid might be incorporated for improved flexibility. A small excess of diol is used, and the acid-catalysed esterification is carried out in the presence of xylene to assist the azeotropic removal of water of condensation. When a degree of polymerisation of about nine is reached, by which time the system is at the higher temperature, it is allowed to cool to below about 140°C, and after loss of xylene, is mixed with the unsaturated monomer, which is generally styrene containing sufficient inhibitor to prevent premature curing. The resin is cooled to room temperature and stored ready for use. The actual method of cure depends upon the specific application, but usually a peroxide initiator is used. For high-temperature cure, say at about 100°C, benzoyl peroxide is commonly employed, whereas for room-temperature cure, as in hand lay-up fabrication techniques, other peroxides with metal salt accelerators are preferred. The reaction is exothermic, that is, it is accompanied by a rise in temperature. Indeed, one of the useful features of inorganic fillers and reinforcing agents in these resins is that they dissipate the heat of reaction more efficiently.

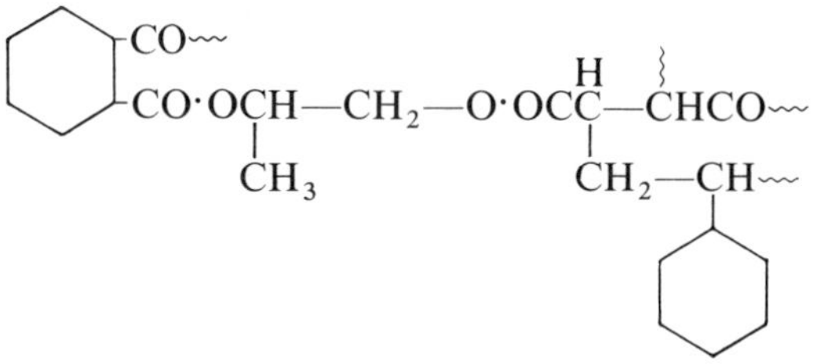

styrenated polyester resin

Like polyester resins, *epoxy resins* have the advantage over many other thermosetting polymers in that, since no volatiles are produced during cure, moulding does not in principle require high-pressure moulding equipment. The epoxies were produced as serious commercial materials in appreciable quantities in the 1950s, some twenty years after the polyesters. The most important resins are based upon bis-phenol A and epichlorhydrin, as shown below. These are typically mixed in a resin kettle at about 50°C and heated slowly to 115°C when, using an alkaline catalyst, a liquid glycidyl ether of molecular weight corresponding to a single ether or a polyether determined by the bis-phenol A:epichlorhydrin ratio in the reaction mixture, is formed. After hot-water extraction of catalyst, the products are dried by distillation when they vary in nature from viscous liquids to solids. Cure of the resins can be effected by a range of compounds, the most important being aliphatic amines and organic acids and anhydrides, with cross-linking taking place between epoxy groups or hydroxy groups. Again the reaction is exothermic.

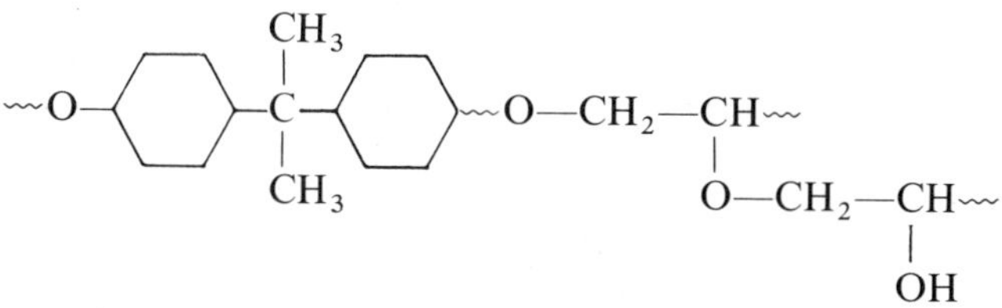

epoxide resin

Other important and related resins are the epoxy novalaks in which the bis-phenol A is replaced by a phenolic resin, and phenoxy resins which are high molecular weight ($\sim 30\,000$), essentially thermoplastic products of epichlorhydrin and bis-phenol A, but retaining a potential for further reaction by cross-linking. Where cross-linking is carried out, an isocyanate, active through the residual hydroxy groups, may be used.

Another group of important thermosetting resins and elastomers, are the *silicone* polymers. These were developed in the USA in the 1940s although much of their chemistry was investigated over a number of years previously in the UK. They are prepared by reaction of an appropriate blend of chlorsilanes in solution, with water, using the knowledge that functionality depends upon degree of substitution. Thus, dichlorsilanes are used for chain extension, trichlorsilanes for branching and cross-linking, and monosubstituted silanes for chain termination. After initial reaction, the system is concentrated and further controlled polymerisation allowed to take place, using a metal octoate. For final resinification the above is applied to a particular situation and cured at elevated temperature in the presence of similar catalysts. Cold curing elastomers (RTVs) are long-chain silicones with a few cross-linking sites which are cured, as above, by means of organo-metallic salts.

$$\begin{array}{ccc}
\mathrm{O} & & \mathrm{CH_3} \\
| & & | \\
\sim\!\!\mathrm{O}\!-\!\mathrm{Si}\!-\!\mathrm{O}\!-\!\mathrm{Si}\!-\!\mathrm{O}\!\sim & \\
| & & | \\
\mathrm{O} & & \mathrm{CH_3}
\end{array}$$

silicone polymer

One of the most versatile classes of polymers presently on the market, which like the silicones can exist as both thermosetting polymers and thermoplastics by correct choice of initial reactants, are the *polyurethanes*. These polymers were first prepared on a commercial scale in Germany in the 1930s, an original interest in them being for use as fibres, to compete with nylon 66. However, it was soon clear that they also form the basis for plastics and elastomers. Like other thermosetting polymers, they find important additional applications as adhesives and coatings, and particularly as foams, both rigid and flexible. Polyurethanes are produced by the reaction of a di-isocyanate and a diol. Between these two and the secondary reactions which are possible, they account for the fact that a wider range of properties and applications are available than for possibly any other polymeric type. Indeed, in the light of the secondary reactions involving the production of allophanates and urea linkages, the term 'urethane polymers', rather than polyurethanes, is sometimes preferred. Two of the more common di-isocyanates are toluene di-isocyanate (TDI) and diphenylmethane 4,4′ di-isocyanate (MDI), and the diols, hydroxy-tipped polyethers or polyesters together with simple glycols or diamines as chain extenders. Small quantities of triols may be used for direct cross-linking. As far as polymer composites are concerned, a range of millable, cast, foamed, and thermoplastic elastomers are of most importance, some being made by a one-shot process in which the reactants are added together all at once, or by a prepolymer route involving two successive stages. If water is present, carbon dioxide is immediately formed on contact with any unreacted isocyanate group, thus providing the basis of foam production.

$$\text{NHCO·O(CH}_2\text{CHO)}_n$$

polyurethane

Before leaving thermosetting polymers, we should have a look at some other lightly cross-linked elastomeric polymers, since these also are commonly compounded with some kind of filler. Historically, the most important example is *natural rubber*, scientifically known and studied since the sixteenth century, but not technologically developed until three centuries later when Mackintosh used the natural latex for waterproofing raincoats, and a little later when the process of sulphur vulcanisation, or cross-linking was discovered. The rubber is mainly isolated by acid coagulation, although about 10 per cent of latex is used directly after purification and concentration. The coagulant is normally either smoked or granulated before baling, when it is ready for compounding and mastication prior to moulding. Of the wide range of *synthetic elastomers* (Table 2.2), the most important is styrene–butadiene rubber (SBR) produced by peroxide catalysed emulsion techniques as distinct from the block copolymer thermoplastic elastomer described later, which is pre-

TABLE 2.2
SYNTHETIC ELASTOMERS

cis 1,4 Polyisoprene	Chlorosulphonated polyethylene
Styrene–butadiene rubber	Neoprene
Butyl rubber	Chloroprene
Silicone rubber	Fluorinated rubbers
Polyurethane rubber	cis 1,4 Polybutadiene
Ethylene–propylene rubber	Thermoplastic elastomers:
Nitrile rubber	Styrene–butadiene styrene block
Vinyl pyridene rubber	Polyether–polyester
Acrylic rubber	Polycarbonate
Polysulphide rubber	Ethylene–propylene propylene block
Polynorborene	Polyurethane

pared by ionic solution techniques as are synthetic cis 1,4 polyisoprene and polybutadiene (containing two parts of isoprene for sulphur cross-linking). Polychloroprene (Neoprene) and nitrile rubbers, the latter based upon butadiene and acrylonitrile, are produced by emulsion polymerisation. A recent emphasis in the field of elastomers is on liquid precursors, although opinion is divided as to their real potential, especially as current interest in this area centres on silicone polymers which tend to be expensive.

Thermoplastic Polymers
The thermoplastic range of polymers constitute by far the greater volume production of synthetic polymers as already mentioned, and of these, five account for the bulk of the total. In addition, in terms of growing importance as reinforcing polymers for engineering applications, competing with metals not only because of their low density but on mechanical property merit, there are a number of step-growth plastics with others of extremely useful properties but disadvantaged on price. For convenience of presentation in this brief review, the class of thermoplastics will be divided into addition polymers and step-growth polymers, the former being described first.

One of the largest tonnage plastics is the *polyethylenes*, distinguished from each other by density and thus reflecting differences in chain branching and crystallinity and in particular, of method of synthesis. Low-density ($0.915–0.94$ Mg/m^3) polyethylene was first produced in 1939 in commercial quantities, using a high-pressure process for the free radical polymerisation of ethylene. (It is interesting to note that ethylene itself was first prepared from molasses, and not from petroleum, a point of some significance at the present time when alternative feedstocks are being investigated as a source of chemicals.) Higher density and therefore more crystalline polymer is prepared using Ziegler catalysts, and even higher density polyethylene (0·96 Mg/m³ and above) also by low-pressure processes using the Phillips and other techniques. The products after recovery are ready for melt processing and compounding. In the related hydrocarbon field of the polyolefines, *polypropylene* is the most important plastic and, of the commodity polymers, is considered to have the greatest growth potential for the next few years. It is made in a similar way to high-density polyethylene using a Ziegler catalyst, the product being separated from the non-crystallisable atactic fraction and catalyst residue, although new processes are available to obviate this stage. The softening point is some twenty degrees above that of high-density polyethylene at about 145°C which gives it an edge for general use. Other polyolefines, not already mentioned, which are of commercial importance are polybutene-1 and polymethyl pentene.

$$\sim\!CH_2\!\cdot\!CH_2\sim \qquad\qquad \sim\!CH_2\!\cdot\!\underset{\displaystyle |}{\overset{\displaystyle CH_3}{CH}}\sim$$

polyethylene   polypropylene

Another important addition polymer is *polyvinyl chloride* with which should be included its copolymers, especially those which contain vinyl acetate and vinylidene chloride. It is prepared by free radical polymerisation mainly through emulsion and suspension techniques, although it is produced also by bulk and solution processes. Similarly the vinyl acetate copolymers and *polyvinyl acetate* itself are made by the same methods, but the preferred method for the former is solution polymerisation and for the homopolymer, emulsion polymerisation, the emulsified product being available immediately for its usual way of processing.

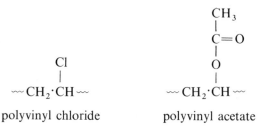

polyvinyl chloride   polyvinyl acetate

Acrylic polymers, of which *polymethyl methacrylate* is the most important, certainly in plastics production terms, cover a wide range of applications, as exemplified by other industrially important representitives, *polyacrylic acid* and *polyacrylonitrile*. The former finds application because of its water solubility, whilst the latter is used as a synthetic fibre and, in fact as such, is one of the precursors for carbon fibre. Polymethyl methacrylate is polymerised by free radical methods, and as well as by the usual four techniques associated with this method as already indicated, it is produced by granulation polymerisation which offers a particularly pure form suitable for use as a moulding powder.

Polystyrene was one of the earliest commercialised addition polymers, like polyvinyl chloride. It is also produced as part of a number of important copolymers as has already been seen, and as will be reported later, is prepared essentially as a polyblend having valuable impact properties far superior to those of the otherwise brittle homopolymer. It is polymerised by a free radical process from styrene using the four major technical processes of bulk, solution, emulsion, and suspension polymerisation as for polymethyl methacrylate. An important large volume production variant is foamed polystyrene in which suspension polymer beads containing trapped blowing agent are heated and moulded in a two-stage steam heating process (the glass-transition temperature of polystyrene is about 100°C) or fed directly to an injection moulding machine or extruder.

The last of the addition class of polymers to which reference is made here, is that based upon *polytetrafluorethylene* and other fluoro polymers distinguished by their low friction and low temperature property advantages. Polytetrafluorethylene is produced from the very pure gaseous monomer by free radical polymerisation, either in granular or dispersion form. Amongst a wide range of other fluoro-polymers, mention is made of important elastomeric copolymers made from vinylidene fluoride and hexafluoropropylene or chlorotrifluorethylene having in addition to other desirable properties, excellent oil-resistance.

$$
\begin{array}{c}
CH_3 \\
| \\
O \\
| \\
C{=}O \\
| \\
{\sim}CH_2{\cdot}C{\sim} \\
| \\
CH_3
\end{array}
$$

polymethyl methacrylate

${\sim}CH_2{\cdot}CH{\sim}$

polystyrene

${\sim}CF_2{\cdot}CF_2{\sim}$

polytetrafluorethylene

Turning to the step-growth class of thermoplastic polymers, historically the most important group of these is that of the *polyamides* of which nylon 66, synthesised by Carothers in 1935, is the best-known example. Other commercialised polyamides include nylons 6, 6.10, 7, 9, 11, and 12. They are prepared by condensation of a diamine and a diacid, self-condensation if an amino acid is the monomer, and by lactam ring opening. As a group they tend to be crystalline and are capable of molecular orientation, so form the basis for fibre formation. The other important application has been as moulding compounds. A relatively new class of polyamides to reach industrial importance are the aromatic polyamides in which the diacid for example, instead of being aliphatic, is aromatic in nature. These are not conveniently prepared by the usual techniques and therefore require interfacial polycondensation methods.

$$\sim OC(CH_2)_4 CON(CH_2)_6 \overset{\overset{\displaystyle H}{|}}{\underset{\overset{|}{\displaystyle H}}{N}} \sim$$

polyamide

Of the saturated *polyesters*, polyethylene terephthalate is the most important, having a much higher melting point than the linear aliphatic polyesters. It is prepared, not by direct reaction of diacid and diol, but for technical convenience by ester interchange of dimethyl terephthalate and ethylene glycol. Although its primary application has been in fibre production by melt spinning, it is also widely used for the preparation of biaxially-oriented films. It also finds some use as a moulding material as does another useful polyester, polytetramethylene terephthalate.

Polycarbonates are the condensation products of diol compounds and carbonic acid derivatives, the most important example being prepared from bisphenol A and a diphenyl carbonate. The preparation of the polymer may be through ester interchange, phosgenation, and interfacial condensation, the first being the most used commercially. The phosgenation process, which involves reaction of bisphenol A with phosgene, produces higher molecular weight polymers than does the ester inter-change method.

Acetal resins based upon polymerisation of formaldehyde are the basis of an important class of engineering plastics. Of the commercially available forms, one is produced by polymerisation of very pure formaldehyde, say in hexane using a triphenyl phosphine catalyst, and the other, a copolymer having a minor component, a cyclic ether such as ethylene oxide.

~OC—⟨ ⟩—CO·O CH$_2$CH$_2$~

linear polyester

~O—⟨ ⟩—C(CH$_3$)(CH$_3$)—⟨ ⟩—O—

polycarbonate

~CH$_2$—O—CH$_2$—O~

polyacetal

Practically all of the polymers so far discussed have been on the market for some time, but in the last ten years or so, there has been an introduction of some newer polymers which though owing their discovery to a stimulus by research into high temperature-resistant and other special-property polymers have been found to have valuable applications as high-performance materials. At the present time they are rather expensive, so cost–performance ratio considerations play a large part in their selection for use, but they appear to be fairly well secured, economically speaking, so that some at least will in the next few years no doubt rise to the status of semi-commodity polymers. Without doubt the most important of these are the *polyimides*, produced typically from a tetrafunctional dianhydride, pyromellitic dianhydride, and an aromatic diamine in a two-stage reaction. The first stage is to the polyamic acid stage when the polymer is still in a condition for further processing, being, for example, soluble in a limited range of highly polar solvents. Subsequent heating of the polyamic acid produces internal condensation to the imide, with its excellent electrical and thermal properties. Reaction of low molecular weight polyamides terminated in amine groups, with pyromellitic dianhydride produces another important class of linear polymer, the polyamide-imides. Ringing the changes of the reaction between tetrafunctional and difunctional chemical starting blocks can produce a whole range of thermally stable, so-called semi-ladder polymers, just as extension of the principle to the reaction of chemical units both having tetrafunctionality, leads to the synthesis of ladder polymers.

Early research in another direction but again with the aim of finding new thermally-resistant polymers, led to the synthesis of the polyphenylenes. However, although stable, they proved to be brittle, insoluble, and infusible and so of little commercial potential. In an attempt to overcome these deficiencies by introducing some flexibility into the benzene-ring chain structure, a series of what might be termed modified polyphenylenes, although they are polymer classes in their own right, have been synthesised. Amongst those which have achieved technological importance are *polyphenylene oxide* especially when blended with polystyrene, *polyparaxylylene, polyphenylene sulphide*, and the *polysulphones*. In principle, in so far as reinforcing agents are concerned in improving the mechanical properties of conventional plastics, the newer polymers already possessing structural advantages can, if the appropriate fabrication procedures are possible, advance the applicational potential of polymers even further. Particularly is this the case where thermal stability is a consideration, a property generally lacking for the replacement of other conventional materials of construction by polymers.

polyimide

FILLERS AND REINFORCING AGENTS[2,3,4]

A wide range of fillers, including pigments and other additives, are used in polymeric systems formulation. Many of these find application in other technologies and their availability owes itself not so much to the demands of the polymer industry, as to for example, paper making, cosmetics, the steel industry, ceramics manufacture, paint making, etc. On the other hand, both the facility of many polymers to accommodate otherwise surplus materials without undue deterioration in properties or even to upgrade behaviour, and their ability to become competitive with other structural materials such as metals, when they are filled with certain fibrous substances, has led to a considerable supplementary demand for new types of fillers. Examples of some of these are shown in Table 2.3.

TABLE 2.3

FILLERS FOR POLYMERS

Particulate		Fibrous	
Organic	Inorganic	Organic	Inorganic
Woodflour	Glass	Cellulose	Whiskers
Cork	Calcium carbonate	Wool	Asbestos
Nutshell	Alumina	Carbon/graphite	Glass
Starch	Beryllium oxide	Aramid fibre	Mineral wool
Polymers	Iron oxide	Nylons	Calcium sulphate
Carbon	Magnesia	Polyester	Potassium titanate
Protein	Magnesium carbonate		Boron
	Titanium dioxide		Alumina
	Zinc oxide		Metals
	Zirconia		Sodium aluminium
	Hydrated alumina		hydroxy carbonate
	Antimony oxide		
	Metal powder		
	Silica		
	Silicates[a]		
	Barium ferrite		
	Barium sulphate		
	Molybdenum disulphide		
	Silicon carbide		
	Potassium titanate		
	Clays		

[a] talc, mica, calcium silicate

The range can be subdivided in various ways, but for convenience this section will group them first of all into particulate and fibrous fillers. Clearly the former will embrace not only fillers of a regular shape such as spheres but also many of irregular shape possibly having extensive convolution and porosity in addition. In general, the aspect ratio, i.e. length:breadth, will be nowhere as high as in the second group, which will include ribbons, sheets, and whiskers as well as the usual continuous filament, staple fibre, and, where metals are concerned, wire. Fillers which can change shape, except by comminution or agglomeration, during composite formation such as occurs with some polymer blends in which both components are fluid during moulding, are left to a later chapter. To help presentation, the fillers and reinforcing agents to be described will be further classified into organic and inorganic.

Particulate Fillers

Organic
Examples of organic particulate fillers which find use in composite polymeric systems include lignin and cellulose-based materials such as both hard and soft wood dust and straw, a range of different nutshell flours, cellulose pulp also from wood, α-cellulose, and ground bark. Starch is another important filler of natural origin, currently finding application as a biodegradable filler for use in disposable plastic bags. Proteinaceous fillers of a similar background have been obtained from soya bean and feathers. In addition, powdered rubber, leather, and plastics, and particularly carbon-black find application. Despite limited thermal stability, organic fillers have an advantage of being of low density, and many have a valuable role as relatively cheap extenders for the more expensive base polymer, as well as providing some incidental property such as reduction of mould-shrinkage, which is of importance in the processing of polymers. Again rather than offering some stiffening or strengthening power which, depending upon the nature of the resin matrix, they may or may not do individually, the more flexible fillers can result in much improved impact strength compared with that of either the plastic itself, or a fibre-reinforced composition. Some of the more important examples of particulate organic fillers will now be described.

Powdered cellulose is obtained from wood pulp associated with the paper-making industry and originates from soft and hard woods as well as from certain grasses. It is almost entirely cellulose in nature and at a microscopic level is actually fibrous. It has the attribute of being of high whiteness and in the past has found particular use in phenolic and related resin systems. *Woodflour* also finds extensive use in the same materials. Obtained from the heartwood of softwood trees such as pine and spruce, the sawdust, chips, etc., are ground in wet stone mills, rollers, or hammer mills. The particles of aspect ratio of about 2·5:1 are sold in grades within the range 50–150 mesh. *Nutshell flour* is an alternative to woodflour but for some applications is mixed with woodflour to produce improved processibility and gloss in the final composite, at the possible expense of stiffness, arising presumably out of the lower aspect ratio. Although a wide range of shells can be used, each type contributing some particular individual property, the bulk of the flour is walnut and coconut.

A wide variety of *starches*, from rice, wheat, maize, and potato has been used as fillers with the grain shape varying between spherical and ellipsoidal with average grain sizes also varying from about 5 microns (rice) to 55 microns (potato). It is found that up to about 10 phr there is no significant effect on composite surface texture or on processing but beyond this there is a definite change.[5] In the use of starch or indeed any filler of an organic nature, it is important to realise that processing, especially of a relatively high melting point polymer, can affect the properties of the filler.

In recent years there has been a growing interest in the use of *powdered rubber* in rubber processing technology, although it is now being admitted that the impact on the industry compared with usual rubber processing, has not been as influential as was thought likely a few years ago. However, it has been used as a filler for plastics for a number of years starting from the time of its original incorporation into polyvinyl chloride in the 1930s. It is made by a number of techniques including cryogenic grinding (when the rubber is below its glass-transition temperature), precipitation, and coagulation. For synthetic elastomers it can be made by direct polymerisation. To prevent the rubber particles adhering to each other and thus forming unwanted agglomerates, additives such as talc or silica are admixed. Before turning to the last of the so-called organic particulate fillers, in connection with the remarks just made with respect to the preparation of powdered synthetic rubber, it should be mentioned that a wide range of emulsion- and suspension-polymerised polymers are available for blending. These have formed the basis for polymer alloys prepared by mixing, say, different emulsions of appropriate type of stabilisation so that premature coagulation does not occur. Similarly step-growth polymers can, in principle, be precipitated from solution to a fine particulate form and so again can be used as fillers.

Particulate *carbon*, quite distinct from carbon or graphite fibre, finds extensive application as a reinforcing agent, processing aid, and as a pigment. By far the main use (>90 per cent) is to be found in tyres as a filler for natural and synthetic rubber. Carbon has another role in producing conductivity in normally electrically insulating polymers, particularly in preventing the build-up of static electricity, and also has a useful incidental role as a UV stabiliser. Although nominally a polycrystalline carbon, its surface, which dictates many of its important properties, can vary not only in porosity and irregularity, thereby potentially increasing interfacial area with a polymer matrix, but in chemical activity. This can involve it in very strong interaction forces, possibly due to unsaturation or due to local oxidation products of which carboxylic, phenolic, quinone, etc., have been reported. Although available as channel, furnace, acetylene, and thermal blacks, in recent years the gas and oil furnace blacks have dominated production. In the preparation of these an aromatic-type oil is fed into a gas or oil flame whereupon it breaks down into carbon black particles which are then quenched in a water spray. Further cooling is used to heat the incoming oil. The particles are typically agglomerated by electrostatic precipitation, variations in this and other steps of the overall process being used to control particle size and size distribution. The high surface activity can lead to aggregation of particles into chains (Fig. 1.3) so that the structure has frequently to be broken down by ball-milling before use as a filler. This does not mean that re-agglomeration will not occur during fabrication, however. Some indication as to the nature of the various carbons is given in Table 2.4.

TABLE 2.4
EXAMPLES OF PARTICULATE CARBON

Type	Code	Type	Code
Channel gas black		Conductive blacks	
Hard processing	HPC	Super conductive furnace	SCF
Medium processing	MPC	Conductive furnace	CF
High colour	HCC	Acetylene	ACET
Medium colour	MCC	Furnace gas blacks	
Low colour	LCC	High modulus	HMF
Medium flow	MFC	Furnace blacks	
Furnace oil blacks		Medium colour	MCF
Super abrasion	SAF	Medium flow	MFF
Intermediate super		Thermal gas black	
abrasion	ESAF	Fine channel	FT
High abrasion	HAF	Medium channel	MT
High structure		Lamp black	LB
High abrasion	HSHAF		

It is of interest to note the code convention for describing the various blacks as HCC, MCC, ISAF, HAF, etc. The particle size and distribution play an important part in the selection of filler for a particular purpose. For ease of processing of a composite, the larger the particle size, within reason, the better because of the relatively low polymer–filler interaction which prevents the viscosity increasing too much. On the other hand, such particles will exhibit inferior blackening propensity. To complicate the selection even further, it is found that some systems of oxidised surfaces behave better with some types of polymers than others.

Inorganic

This class of filler constitutes probably the more important group of particulate fillers in view of their low price levels and ready availability, thus immediately providing a basis for reducing the cost of moulded articles without too much loss, if any, of accepted properties. Indeed, they also provide in some cases a welcome improvement of properties, for example in rigidity, or offer some complementary advantage in mixed systems of particles and fibres. Table 2.3 showed the range of inorganic particulate fillers, but reference here will, again, be only to the more important examples.

Silica, which is the most abundant mineral on earth, occurs naturally in various degrees of purity as well as in combination with other inorganic groups, as in silicates. In addition some important fillers are produced synthetically. Ground silica is obtained from quartz sands which are given a primary crushing, washed, and then dried. Pulverised silica is obtained by further grinding in a ball mill and is then graded by screening and air flotation. A microcrystalline form of silica and hydrated silica occurs naturally as diatomaceous earth in the form of Kieselguhr, Tripoli, etc. It has a porous structure derived from its parent fossilised diatoms and is crushed if necessary before being graded. Forms of so-called amorphous silica can be made synthetically by controlled acid precipitation of aqueous sodium silicate, by electric-arc vaporisation (thermal silicas) and, especially, by burning mixtures of silicon tetra-chloride vapour and hydrogen or natural gas in air (fumed or pyrogenic silica). Fused silica is prepared from high quality quartz by fusing in arc furnaces, following which it is ground. It should be mentioned that, as for the case of the furnace blacks for carbon, particle size of the 'amorphous' silicas can be controlled by attention to processing details (Table 2.5).

TABLE 2.5
SILICA FILLERS

Natural	Synthetic
Silica sand	Precipitated
Quartz	Pyrogenic
Novaculite	Aerogel
Diatomaceous earth:	Hydrogel
Kieselguhr	Silica fibre
Tripoli	Quartz fibre
	Pulverised silica
	Fused silica

The most important silicate used as a filler is hydrated magnesium silicate (talc) which exists naturally in many forms and like other minerals is ground after purification, and then seived. The other important silicate is calcium metasilicate (Wollastonite) available in a range of grades.

Moving to the next group of mineral fillers, the *clays*, it is found that the particles of these fillers, derived from the weathering of feldspar rocks, are frequently plate-like in shape. Chemically, they are largely mixed aluminium and silicon oxides with substitution of other metal oxides such as magnesium, iron, and potassium. They exist in Nature as kaolin, mica, vermiculite, etc., and usually contain other mineral impurities. Kaolin, which is essentially a hydrated aluminosilicate, perhaps better known as china clay, is obtained by open mining, then washed and decolorised in aqueous suspension before drying. Anhydrous grades or calcined clays are prepared by heating to about 600°C and then milled before use. Mica occurs naturally as muscovite, phlogopite, biotite, etc., the chemical nature being slightly different in different micas but all are characterised by their ability to cleave into sheets of about 25 microns or less, and thus in a planar sense, exhibit high aspect ratios. They are supplied as fillers after dry or wet grinding, or in some locations, by both. Vermiculite, which coexists with common mica, undergoes a large expansion on heating as bound water is lost. It is then in a very friable condition so that it can be ground into a particulate state.

In addition to the mixed oxides mentioned above, a number of simple *metal oxides* also find application as fillers. One such oxide is alumina, which is found in the hydrated form as bauxite. In one process it is ground, calcined, and then further ground before being added to caustic soda solution to form sodium aluminate, the major impurities being insoluble. After cooling the solution is seeded with hydrated alumina which precipitates the same material. This is washed and heated at 1000–1200°C, when it is ready as a filler, although the bulk of the material is actually used in aluminium metal production. Titanium dioxide is a very important filler, but is mainly incorporated as a very effective white pigment, as will be seen later. It occurs as rutile and ilmenite, which in purification are treated with sulphuric acid, hydrolysed, and then calcined back to the dioxide. Alternatively, rutile is converted to titanium tetrachloride by treatment with chlorine and then oxidised, or hydrolysed with steam, to titanium dioxide. Further high temperature treatment results in allotropic changes. Other oxides which find application mainly as pigments or special purpose fillers include beryllium, lead, iron, magnesium, zirconium, and, particularly, zinc oxide.

Calcium carbonate is not only the most important carbonate used as a filler but it is also the major particulate filler, finding its main outlet with polyvinyl chloride composites. It is obtained by the quarrying of chalk, limestone, and marble, and is then coarse ground, screened, and ball-milled. It may be graded by elutriation, the finest grades having the least tendency to settle, dry milled material being found in this way to be coarser than wet milled. The finest particle sizes for calcium carbonate are obtained through precipitation techniques. In one such method, calcium carbonate in a finely divided form (approximately 0·05–10 microns) arises as a by-product of the Solvay process for the production of sodium carbonate. Other approaches are the sodium carbonate precipitation of slaked lime or the corresponding precipitation brought about by carbon dioxide. The other carbonate which should be mentioned is the mixed carbonate of calcium and magnesium, which is found as the minerals dolomite and magnesite.

In addition to a number of specialised salts such as calcium sulphate used with pigments, molybdenum disulphide for lubrication, barium ferrite for magnetic properties, barium sulphate for chemical resistance and X-ray opacity, there are a few derived materials used as fillers. The first group of these is that of the *metal powders*. These again are used for some specific purpose, e.g., metallising effects, increased electrical and thermal conduction and (with lead) high density and radiation resistance, including the special case of sound absorption. Practically all metals can be prepared in powder form by reduction of oxides, decomposition of carbonyls, electrochemical displacement, and electrolytic decomposition. The list of metals used as fillers includes aluminium, copper, bronze, brass, iron, silver, steel, and zinc.[6]

The second group of derived inorganic particulate fillers is exemplified by *glass*. Glass, in somewhat different chemical compositions, is produced in many forms so that although fibres are the most important glass fillers for polymers, it can be produced and indeed, is available, in many other geometries. These include powder, beads as ballotini, flake and hollow spheres, and in a related form as a by-product of the electrical power industry as fly ash or hollow cenospheres. A use of the hollow grades is for the preparation of syntactic foam composites. Solid spheres can be produced by dropping powdered glass, usually A-type glass, through a gas flame, or alternatively molten glass may be dispersed by blowing air over the melt. The size distribution is fairly large, 50-micron beads coated with a coupling agent being the preferred size. Hollow spheres are made by melting glass particles containing a blowing agent, the final size being about 20–130 microns and having a narrower distribution than do cenospheres. Glass flakes used as reinforcing agents are available in various thicknesses and are prepared, for example, by crushing fine tubing made from, typically, E or C glass.

Some mention has already been made of the use of pigments in polymers, and to the fact that they may be added at levels comparable to those associated with filled and reinforced plastics. This being the case, although not of prime concern to this text, the general nature of pigments should be recognised, and for this purpose a list of some better known examples is provided (Table 2.6).

TABLE 2.6
PIGMENTS FOR POLYMERS

Inorganic	Organic
Titanium dioxide	Carbon blacks
Iron oxides	Phthalocyanines
Lead chromate/molybdate	Mono azo
Cobalt aluminate	Metal toners
Nickel titanate	Diarylide yellows
Chromium oxide	Pyrazolenes
Ultramarine pigments	Azo condensates
Manganese violet	Vat pigments
Iron blue	Iso indolinones
Copper chromite	Quinacridones
Cadmium pigments	Flaventhrones

Fibrous Fillers

Organic

Although at the present time for reasons of inherent strength, stiffness, thermal stability, and sometimes cost, inorganic fibres, particularly glass fibres, are the most important fibre reinforcements for polymers, recent developments in high modulus and thermally-resistant organic fibres are creating interest, especially where weight considerations come into the decision making, as for aerospace applications. Some have a useful sacrificial role in ablative plastics composites where, being organic, they produce volatile degradation products which cool the eroding substrate.

Of the natural fibres, *cellulosics* are the most important and include the short length fibres rejected in the cotton spinning process, appearing on the market as flock. They are preferred to chopped cloth and yarn, as well as the corresponding products of regenerated viscose fibres. In the woven state, cellulosic fabrics are often used as laminating material. The ligno–cellulosic fibres such as jute and sisal have had some application in phenolic and polyester systems. The other natural fibres based on proteins, wool, and feather, have found some small application, with wool being used as a bodying agent.

Over the years many synthetic fibres have been produced, covered by the main classes of nylons, aromatic polyesters, acrylics, polyolefines and polyethylene, polyurethane, and polyvinylidene chloride. There are many ways by which synthetic fibres can be made, but the main ways are by spinning from the melt, solution spinning, and dry spinning. The first method is particularly useful for thermoplastic polymers, providing the melt viscosity is not too high and the melting point also is not so high that degradation takes place. It further offers scope for producing fibres of different cross-sectional profiles by shaping the spinning die, although circular cross-sections are by far the most common. In wet spinning, a solution of fibre-forming polymer is extruded into a coagulating bath, whilst in dry spinning the polymer is spun in a volatile solvent which evaporates on extrusion from the die. All techniques so far mentioned produce continuous filament which may be used directly, or may be chopped to predetermined length, possibly with crimping, to give staple fibre products. The processes also afford facilities for variation which in turn leads to some control of final fibre properties. Typical of this control is through variation in speed of extrusion and the amount of stretch, characterised by the draw ratio, given to the extruded filament. In recent years considerable emphasis on this aspect has enabled the production of *ultra-high modulus fibres*, in particular from polyethylene. Using a combination of tensile drawing and hydrostatic extrusion, fibres of draw ratios of about $\times 40$, i.e. about ten times those of conventional textile fibres, have been achieved. These fibres are extremely stiff compared with ordinary fibres (Fig. 2.2) having moduli in the range of 50–100 GPa, again about ten times those of textile fibres, and approaching something like a third or more of theoretical maximum stiffness. These developments are of great potential for composites offering as they do high stiffness with low weight. Of the ordinary fibres, nylon has found application in phenolic ablative plastics, and in what is really a different context, polypropylene fibres have been used to reinforce inorganic cement. Undoubtedly the most interesting organic fibre class for use in polymer composites at the moment is that of the aramids, or aromatic

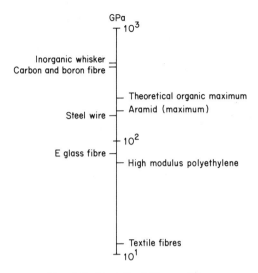

FIG. 2.2. Moduli of fibrous fillers.

polyamides, although some report has been made of the use of polyester fibre as well.[4]

The process leading to the production of *aramid fibres* is shrouded to some extent in commercial secrecy, but they are believed to be manufactured by interfacial or solution polycondensation methods. They do not show melting in the conventional sense, although crystalline, because of the very high viscosity of the polymer, arising from the intrinsically stiff chains based upon aromatic precursors. In fact, because of this stiffness, not only have they to be spun from very polar solvents, such as pure sulphuric acid, but the molecules automatically align under shear, giving rise to liquid-crystal behaviour. Of the fibres which have been introduced since about 1972, the best known at the present time is 'Kevlar 49' and the slightly lower modulus fibre, 'Kevlar 29'. The fibres are available as woven and non-woven fabrics, fibres, and rovings. Some difficulty was experienced in producing chopped fibre because of resistance to cutting. It is of some significance that aramid fibres of modulus as high as 130 GPa have been reported.[7]

Although *carbon fibres* have been known for over 100 years it was only in the 1950s that there was a revival of interest in a completely new context as high-strength, light-weight reinforcements which might be of application in aerospace components and which could be used as reinforcements for metals as well as plastics. The fibres can be produced from a variety of starting materials, but at the present time, acrylic fibres, cellulosic rayon, and mesophase pitch are the most important. In the rayon process, rayon filaments are given a chemical treatment with, say, zinc chloride or ammonium chloride, following which they undergo relatively low temperature oxidation. The temperature is then taken up in two stages to about 1000°C, i.e. to the region of the unstable triple point, when graphitisation takes place, the fibre at the same time being hot-stretched to increase modulus. In the case of the acrylic route, following early work on polyacrylonitrile fibre which indicated a higher yield of carbon fibre than did cellulosic fibre, acrylic copolymer fibre precursors were then used. These are oxidised, again first of all at low temperatures (200–250°C), carbonised at about 1000°C, and graphitised at around 2000°C in an electric arc furnace. In addition to single fibres, both graphite and carbon woven fabric forms are available. The only difference here, compared with other textile fabrics, is that the conversion is carried out on previously woven material, instead of the fabric being woven from the fibres themselves.

Inorganic Fibres

As for organic fibres, these are produced in both natural and synthetic forms, the most important of the former being asbestos, a material which has come in for considerable scrutiny in the light of associated health hazards. Glass is without doubt the most important of the latter although quite a number of other synthetic inorganic fibres have now reached the market. Others are still in the experimental stage, where possible applicational roles for the reinforcement of metals as well as plastics is also of interest.

Asbestos occurs naturally as two major classes, serpentine and amphibole, in the form of a number of hydrated magnesium silicates, the amphiboles also containing other metal elements. It has been used as a reinforcing agent for polymers in a variety of forms—fibre, yarn, and cloth—because of its relative cheapness and stability. To what extent health reservations will affect its future is not yet fully appreciated. The most important form of asbestos, amounting to about 95 per cent of total production, is the serpentine fibre, chrysotile, produced in different countries notably Canada and the USSR. The fibre lengths of asbestos vary with the extraction history, which is completed by dry screening, but is typically about 25 mm and even this length may be further reduced by the compounding and moulding procedures of composite production. The filler shows compatibility with a wide range of, particularly, polar polymers, and is capable of forming high-tensile-strength materials. Although it is generally classed with particulate fillers, perhaps because of its fine nature, it is strictly fibrous as already indicated, having an aspect ratio of about 15:1.

Of what might be termed the synthetic inorganic fibres, the most important in view of its scale of usage is *glass fibre* and, indeed, until very recently, to many people the use of the phrase 'reinforced plastics' meant essentially glass-reinforced plastics (GRPs). The fibre can be produced in different ways but the main process is the direct melt process which is superseding the glass marble route (Fig. 2.3). In this, the molten glass is

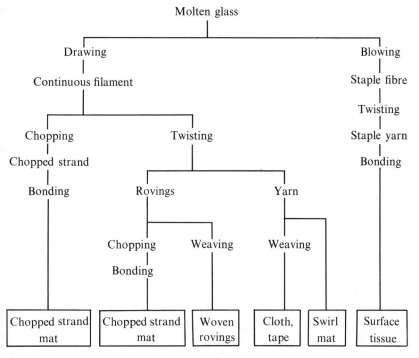

Fig. 2.3. Glass fibre manufacture.

extruded through several hundred spinneret holes, through a sizing bath to protect the integrity of the fibres which are very friable, the bath also possibly containing coating agent. The strand of filaments is now converted to roving by bringing together a number of strands, or is cut to make chopped strand. The filaments are typically 9–18 microns in diameter, although finer deniers are made as well. The glass used is designated in a number of ways, such as 'A' (soda glass), 'E' (electrical grade), 'C' (chemical resistant), and 'S' (high strength), but other grades, e.g. 'D' and 'R' grades are manufactured. At the present time 'E' glass is the most important for general use. Chopped strand is used directly for reinforcement or is used to form chopped strand mat (CSM), the fibre lengths of which are typically 3, 6, and 12 mm. The continuous filament is used in filament winding and pultrusion techniques of composite manufacture, and rovings are used to prepare woven fabrics, etc., as are also strands. Various textile parameters relating to twist and to different types of weave are adopted to provide a wide range of products. Other arrangements such as tapes and swirl mats are available and at the other extreme, strands may be hammer-milled to produce hammer-milled fibre of length about 3 mm.

About the time of the developments which were taking place in carbon fibres similar activity was taking place, particularly in the USA on *boron fibres*. These fibres of high strength and low density have been used, for example, in the reinforcement of epoxy resins. They are prepared in different ways, but the conventional method is by deposition of boron from a boron trichloride–hydrogen system on to a fine tungsten wire, of about 10^{-3} cm in diameter, heated electrically to about $1300°C$, to make a two-component concentric filament of just over 10^{-2} cm in diameter. It can then be impregnated with an epoxy resin to form a prepreg material ready for polymer composite manufacture.

To conclude this section on inorganic fibres, mention is finally made of a wide range of miscellaneous fibre types which have found application in polymer composites but probably not on the scale of any of the above examples. The first group of these is the *ceramic fibres*, in particular those based upon alumina, at least one commercial product having as coating a film of silica to increase strength through the reduction of surface flaws, and at the same time to increase compatibility with molten metals, for which, in addition to their use as possible reinforcing agents for polymers, they have been developed. It is claimed that because of their polycrystalline nature they have better transverse properties than aramid or graphite fibres. They are available in different fibre lengths and as fabrics. Other ceramics which have been produced in fibre form include silicon nitride and silicon carbide as well as high silica and quartz.[8] *Metal filaments* in the form of fibres, wires, and whiskers have found a limited application in polymer composites. The fibres are available in a range of cross-sectional geometries and in different materials including aluminium, copper, steel, and tungsten and are made by drawing of wire, melt extrusion, foil slitting, etc. They are put on the market also as felts and fabrics. The last class of fibrous filler to which attention is drawn is that of *whiskers*. These are single-crystal materials of very high strength and, compared with the corresponding fibres, high elastic strains. They have been made from many materials, ceramics and metals, typically by high-temperature vapour-transport reactions, probably the best known whisker materials being silicon carbide, silicon nitride, alumina (sapphire), aluminium nitride, etc. Although short in length, of the order of microns, they have very high aspect ratios of many hundreds. At the moment, despite their impressive property potential as far as composites are concerned, the cost of production does not allow them to compete with conventional reinforcing agents.

REFERENCES

1. BRYDSON, J. A., *Plastics Materials* (3rd Ed), Newnes-Butterworth, London (1975).
2. KATZ, S. H. and MILEWSKI, J. V. (Eds), *Handbook of Fillers and Reinforcements for Plastics*, Van Nostrand-Reinhold, New York (1978).
3. WAKE, W. C. (Ed), *Fillers for Plastics*, Iliffe, London (1971).
4. SEYMOUR, R. B. (Ed), *Additives for Plastics*, Academic Press, New York (1978).
5. GRIFFIN, G. J. L., in Deanin, R. D. and Scholts, N. R. (Eds), *Fillers and Reinforcements for Plastics*, ACS, Washington (1974).
6. SHELDON, R. P., Chapter 4 in ref. 3.
7. See DOBB, M. G., JOHNSON, D. S. and SAVILLE, B. P., *Phil. Trans. Roy. Soc. Lond.* **A294**, 483 (1979).
8. GALASSO, F. S., *High Modulus Fibres and Composites*, Gordon and Breach, New York (1969).

Chapter 3

MECHANICAL PROPERTIES OF COMPOSITES

MECHANICAL PROPERTIES OF UNFILLED SYSTEMS

In many ways the mechanical properties of polymers are probably the most important of all their properties, since whatever may be the reason for the choice of a particular polymer for some application, whether it be on thermal, electrical or even aesthetic grounds, it must still have certain characteristics of shape rigidity and strength, for example, for it to support at least its own weight. For the special case of composite polymers, the mechanical properties invariably assume a dominant role since so often in these one is seeking an improvement in mechanical behaviour as a prime requirement. Since, by definition, the polymer constitutes the continuous phase, the filler acts essentially through a modification of the intrinsic mechanical properties of the polymer, with such factors as concentration, type, shape, and geometrical spatial arrangement of the filler within the matrix contributing to the detail of this modification. This being the case, in order to appreciate the part played by the filler, it is useful at this stage to consider briefly the nature of the mechanical properties of the unfilled polymer itself.

As a class of structural materials, polymers as a whole are not only very variable in the way that they respond to the application of mechanical forces, but for many of their intended uses, particularly those requiring some stiffness and resistance to fracture, they are frequently regarded as being inferior and thus second rate when compared with other materials such as metals. In the past this condemnation has been generally because of a failure in design, the basic mechanical property data receiving inadequate recognition. For the same reason, because they possess features foreign to the experience of many engineers, the reputation of some of the emerging polymer composites, could easily be similarly damaged. Especially could this be the case where an apparent improvement in properties, or at least no serious deterioration in properties, brought about by the use of inexpensive filler, is reported. What may be overlooked, if assessment is made, say, purely on stress–strain behaviour, is that some other property, critical to the broader use of the material, might result in premature or even catastrophic failure. Returning to a consideration of the mechanical properties of unfilled polymers, let us see how the adoption of standard test procedures for structural materials can lead to results which require very careful analysis.

The most common way of recording mechanical properties is to carry out stress–strain, or more precisely load–extension, measurements using some kind of tensile testing equipment operating at strain rates usually approximating to those suitable for most metals. Indeed, on this basis it was shown a number of years ago that all polymers could be assigned to one of five different classes.[1] These are shown in Fig. 3.1. However, use of

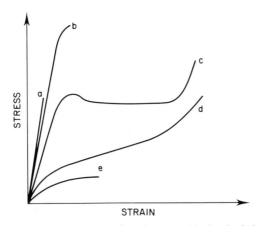

FIG. 3.1. Stress–strain behaviour of polymers: (a) hard, brittle; (b) hard, strong; (c) hard, tough; (d) soft, tough; (e) soft, weak.

this classification soon showed that the overt stress–strain behaviour was very sensitive to a large number of variables, many of which were capable of exerting a significant influence and yet had not been fully recognised by the person making use of the data. This perhaps emphasises the continuing need for co-operation between engineer and scientist at a time when so many new materials systems are being developed. One of the first parameters to be defined is that of molecular weight, reflecting in many cases the method of polymerisation of the polymer and its origin. The lower the average molecular weight, the softer invariably is the polymer. Then again any branching or cross-linking will further influence properties. The presence of impurities or low molecular weight additives such as, for example, moisture or organic liquid, will for the most part also produce a softening, as well as a weakening, effect. One of the most serious influences which can affect properties is temperature. In fact it is not exceptional for a polymer to traverse all five of the above classes of mechanical properties in a temperature range of, say, no more than a hundred degrees. Thus, a polymer like polyvinyl acetate might very well be of use for the construction of buildings in a polar environment and yet merit application only as a flexible contact adhesive at the equator. For an amorphous polymer, the biggest transition in mechanical properties takes place at the glass-transition temperature, corresponding to about a thousand-fold change in stiffness. Any cross-linking between polymer chains will push up the transition temperature and where this is very extensive, as in say, phenolic resins, polymer degradation might intervene before this fall in modulus can occur. Crystallinity in the polymer presents an intermediate case, with some of the rigidity being retained through the stabilising effect of the crystalline regions, which themselves only fail when the melting point is reached. It does not take a great deal of imagination, on this basis, to presume the general influence of a filler, particularly if it exercises some specific interaction with otherwise freely moving polymer chains.

Another factor which affects the response of a polymer to strain, is the rate at which the strain is effected. The general influence of this is felt most strongly at temperatures just above the glass-transition temperature when an increase of deformation rate causes an increase in apparent modulus and also usually gives rise to a more brittle-like failure of the polymer. The combined effect of temperature and strain rate highlights a property characteristic of most polymers in the normal range of stresses and strains experienced by them in their day-to-day applications. This is their viscoelastic nature which arises, at least for amorphous polymers, because of the finite natural relaxation rate of polymer chains, this itself being a varying quantity relevant to the spectrum of chain movements within a particular polymer at a given temperature. For crystalline polymers, where the overall relaxation behaviour is affected by the restricted movement of those chains arranged in crystalline regions, a new type of time-dependent response can arise through structural slippage within these regions. The viscoelastic character is commonly observed in a number of ways. The first of these is creep, in which a polymer sample undergoes time-dependent extension at a constant applied load level: another is stress relaxation in which a deformed specimen exhibits a fall in stress without any change in strain, and a third example of viscoelastic behaviour is in the dissipation of energy as heat in cyclically stressed material, the phenomena being known as hysteresis. This can be experimentally recorded as an energy loss peak occurring at some thermal transition, such as the glass-transition temperature, when the loss is expressed as a function of temperature. In the light of its relationship to molecular movement, it may also be studied as a function of applied cycle frequency. The experimental approach concerned with energy loss indicated above is known as dynamic thermal mechanical analysis. Since fillers can have interaction with polymer chains, it is not surprising that this technique, which reflects molecular movement, is often used in studies of polymer–filler interaction.

Having considered deformation behaviour of polymers let us now have a look at the failure behaviour. There is an immediate problem here since there are different degrees to which the description of failure is given. For example, for many useful applications of a polymer, or for that matter any structural material, it will be considered to have failed when the yield point is reached since this gives rise to permanent deformation. Thus, expressed in simple terms, the polymer is not what it was in a mechanical sense. The yield, however, is not necessarily a condition of failure, as is illustrated by the process of cold-drawing. In this, often in partially crystalline fibre-forming polymers, yielding is accompanied by molecular orientation, possibly with overtones of crystal reorganization, which continues without appreciable change in the stress level, until complete. We now have essentially the same polymer but with a more anisotropic property–structure relationship. In fact, the modulus of the cold-drawn material is much greater than that of the previously undrawn polymer. Eventually fracture takes place, but the approach to final fracture also varies from one polymer to another and from one set of experimental conditions to another. However, before going into this, it is useful to have a look at the fracture process in a little detail.

Fracture may be essentially brittle or ductile. The basis of the theory of brittle fracture was established by Griffith when he postulated that fracture takes place by crack growth.[2] On the basis of an energy balance relating to decrease of elastic energy at the expense of an increase in surface energy during crack growth, he was able to derive the following equation:

$$\sigma_s = \sqrt{\frac{2\gamma E}{\pi c}} \tag{3.1}$$

where σ_s is tensile strength, γ is surface energy/unit area, E is Young's modulus and c is crack length. When measurements were carried out on the fracture of plastics, it was frequently found that as for other materials, brittle plastics apparently obeyed this equation, at least in form, but that the calculated surface energies were much higher than were originally expected from the chemical nature of the particular polymer. Detailed study of the crack environment soon showed that fracture is accompanied for these plastics by a local deformation process involving plastic flow and molecular orientation. In addition there was often evidence of a phenomenon known as crazing, so that the whole behaviour was associated with much greater energy requirements than would be needed for a simple crack formation of a new surface. Crazing is the name given to the formation of what appear to be forms of microcracks containing oriented polymer chains; they are frequently associated with, and are precursors to the formation of, cracks (Fig 3.2).

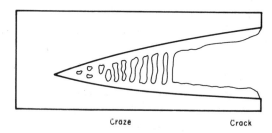

Craze   Crack

FIG. 3.2. Crack preceded by crazing in polymer failure.

They are of interest in another connection, which will later be seen to be relevant to polymer composites, in that unlike single cracks they are able to withstand a certain amount of load stress, and give rise to opacity in the same way that fillers also do in many composites. Indeed, the presence of filler can influence the formation of crazes by modifying force fields, in a way that would not happen in the absence of filler. Two other aspects of fracture should be noted. The first is connected with impact behaviour of polymers in which a material is subjected to high rates of strain which results in complex multiaxial stressing, and which for reasons elaborated above can give rise to brittle-type fracture where otherwise, at lower rates of strain, the material may appear to be tough and non-brittle. The tendency for brittle fracture under impact conditions is increased where, perhaps because of geometrical design of a particular article or because of an incidental or deliberately introduced notch (in essence being a crack), some form of stress concentrator is present. Temperature also will affect the way in which a polymer fractures under impact since the impact strength of a polymer increases sharply in many cases at some temperature just above the glass-transition temperature. The aspect of stress concentration, again because of force field pertubation, can be of some importance in the case of composite systems, so that impact strength is, in fact, very sensitive to the physical and geometrical nature of the filler. The second aspect of fracture which is peculiar to polymeric systems is that which arises from early fracture brought about by certain liquids, the nature of the active liquids depending to some extent on the nature of the polymer. It is usually identified with the presence of inherent stresses in a fabricated material, arising from the nature of the moulding process and is known technically as environmental-stress-cracking.

A property connected with fracture but indicative of progressive failure in materials is that of fatigue, and is of special concern in the use of engineering plastics. It is found that many materials, not just polymers, when subjected to repeated stressing can fail at stress levels lower than those obtained in continuous static loading. This behaviour, termed fatigue, is usually studied by cyclic loading experiments, the number of cycles to failure, for a given level of stress, being recorded. It can also occur as static fatigue where a sample has undergone long-term loading. Again because of the way that fillers can influence stress fields in composites, this property is important to our later considerations.

Amongst the miscellaneous mechanical properties of polymers which have a relevance to composite systems are those of friction and wear. Some indication of the nature of these will be given when the friction and wear properties of composites is considered later. Perhaps the only other property to which attention should be drawn at this stage is Poisson's ratio, which is a measure of overall volume change of a material subjected to a deforming force, and as such makes a contribution to a number of equations relating to composite mechanical properties. The value of Poisson's ratio varies from system to system, but typically for a soft elastomer, having something of the attribute of a liquid in which no volume change occurs on deformation, it assumes the value of 0·5. For less fluid but still flexible polymers the value is nearer to 0·4 and for rigid polymers and fillers, a value of about 0·3 is usual.

MECHANICAL PROPERTIES OF PARTICULATE-FILLED POLYMERS

The mechanical properties of fibre-reinforced composites are perhaps of more interest from the point of view of structural applications, but before discussing these we should have a look at the mechanical properties of particulate-filled polymers. The reason for this is, that although the primary role of the filler may be as an inexpensive extender, pigment or, say UV stabiliser, the ultimate composite must still have suitable mechanical properties for some intended application.

Although it might be thought that the influence of a particulate filler on the deformability of a polymer may be purely hydrodynamic, that is, it merely distorts the molecular flow pattern in the deforming polymer, in practice specific interaction very often produces an enhanced stiffening effect. In the case of carbon-filled rubber where it is believed that a filler particle is surrounded by a sheath of immobilised rubber molecules, the loss in overall polymer mobility reflected in stiffening, and separately in strength, can be quite considerable.[3] The micromechanical analysis of the mechanical behaviour in terms of the separate contributions of the two components to the mechanical properties is complicated. This complication arises not only from fairly straightforwardly recognised complexities of filler concentration, but also from uncertainties in the magnitude of interaction, especially as this might itself vary as the polymer and filler are mechanically forced into greater contact during deformation. In addition there are uncertainties in filler size distribution, complicated by any agglomeration, and to what extent void formation has occurred during fabrication, and the related problem of imperfect interfacial contact between the matrix and the filler. Nevertheless, theories to describe the mechanical properties of particulate-filled polymers have been developed, with often one theory applying to one situation better than another.

Thus, for example, for soft or elastomeric matrixes, the influence of filler may be regarded as being one of some proportional effect on viscosity and so can be represented by Einstein's equation for dilute suspensions:[4]

$$\eta = \eta_1 (1 + k_E v_2) \qquad (3.2)$$

where η and η_1 are the viscosities of filled and unfilled polymer, k_E is Einstein's coefficient ($=2\cdot5$ for spherical particles), and v_2 is the volume fraction of the filler. Amongst other equations, one due to Mooney[5] has been found useful for describing the influence of more concentrated systems of elastomers. This can be expressed as:

$$\ln G/G_1 = \ln \eta/\eta_1 = k_E v_2/(1 - v_2/v_m) \tag{3.3}$$

where G and G_1 are the shear moduli of the filled and unfilled polymer, and v_m is the maximum packing fraction ($=0\cdot74$ for hexagonal close packing and $0\cdot524$ for simple cubic packing and so on). For the special case of the matrix and filler respectively taking on values of Poisson's ratios of $0\cdot5$ and $0\cdot25$, Van der Poel has reported the following equation[6] which has found some application:

$$G/G_1 = 1 + [1\cdot25v_2/(1 - 1\cdot28v_2)]^2 \tag{3.4}$$

Turning to rigid matrixes appropriate to, say, plastics rather than elastomers, i.e. where Poisson's ratio is less than $0\cdot5$ and where the moduli of polymer and filler are closer together in value, the stiffness behaviour can be represented by an equation due to Kerner.[7] In a modified form, due to Nielsen and coworkers,[8] it may be written as follows:

$$G/G_1 = \frac{1 + ABv_2}{1 - B\chi v_2} \tag{3.5}$$

where $A = k_E - 1$, with k_E assuming different values appropriate to the shape of the filler,

$$B = \frac{G_2/G_1 - 1}{G_2/G_1 + A} \quad G_2 \text{ being the filler modulus,}$$

and

$$\chi = 1 + v_2(1/v_m^2 - 1/v_m)$$

The form of the equation is applicable to Young's modulus as well. An equation which takes into account adhesion between the two components and is frequently quoted is that of Hashin and Shtrikman.[9] Incidentally, that there is interaction between particulate filler and polymer matrix has been indicated by the development of asymmetry in mechanical loss curves in dynamic thermal mechanical analysis studies.[10]

Examining now actual data for composites, Fig. 3.3(a) shows the stress–strain curves for a filled elastomer, whilst Fig. 3.3(b), (c) show typical data for a particulate-filled composite having a rigid matrix. In both

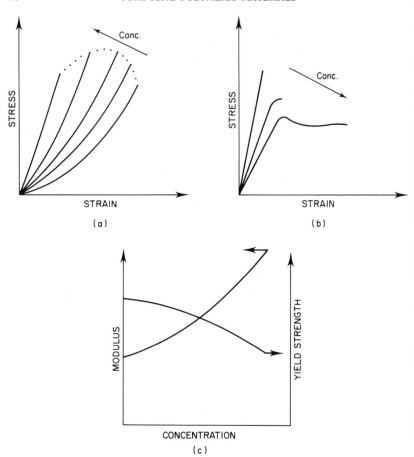

FIG. 3.3. Stress–strain behaviour of polymer–particulate filler composites. (a) soft matrix, hard filler; (b) hard matrix, soft filler; (c) hard matrix, hard filler.

cases, as expected, the modulus increases with increased filler loading. However, this may not always be the case, since if fabrication is accompanied by extensive void formation, the modulus of the actual composite may decrease. Other factors, not described above, which can further modify behaviour include particle size and particle size distribution: the smaller the particle size the greater the increase in modulus, perhaps arising from a greater total surface interaction or from a change in the value of the maximum packing fraction. A further modification will arise where, as mentioned for rubber, a strong local adsorption of polymer chains introduces effectively a third phase. Agglomeration of particles, arising from strong particle–particle interaction, might dissociate trapped polymer from the main polymer matrix and in so doing produce an effect of increased stress for a given strain, i.e. increased modulus. The further influence of particle shape is discussed later in connection with fibrous fillers, but suffice it to say at this point that asymmetry in shape tends to increase modulus.

It will have been noticed in Fig. 3.3 that although the modulus increases for both soft and rigid matrixes, tensile strength and elongation at break do not follow the same relationship. Indeed, although tensile strength, particularly for a rigid matrix and rigid filler, will decrease with increase of filler (attributed to an increased stress concentration effect as well as through the formation of microcracks either at the interface or locally, in the matrix), in the case of a soft elastomeric-based composite which is capable of dispersing stress more effectively, the tensile strength may very well increase. There will of course be some maximum to be reached in the latter case, if for no other reason than that eventually the matrix continuity will be replaced by particle–particle contact. Thus mechanical coherence, except for some agglomeration, will disappear. On the latter point, it follows that improved dispersion can have a contrary effect on strength compared with modulus. Considering the other extreme property, elongation at break, it might be expected that this quantity will fall with increasing concentration, since proportionally, more and more of the experimental strain is being provided by intrinsically less polymer, and in fact, this is usually the case. However, the use of soft fillers in a rigid matrix can give rise to an increase in elongation at break and also in impact strength. Part of this may derive from the ability of some fillers to promote craze formation in deformed polymers prior to fracture. Other factors which affect ultimate tensile properties are, as before, interfacial interaction, particle size, and geometrical shape, but this time having more proportional influence than on modulus. Large particles give rise to greater stress concentration and thus lower tensile strength, than do smaller particles. Where bonding is weak, then at some critical strain debonding takes place and the composite exhibits opacity. But where a suitable bonding agent has been employed, a greater level of stress will be required to produce breakdown of interfacial adhesion. In fact, if the interaction is extremely strong, fracture of matrix or even filler may occur first. In this context it has been shown that for a calcium carbonate-filled polyethylene composite, use of 1 per cent titanate coupling agent can give rise to an approximate increase in tensile strength of 20 per cent.[11]

Turning now to a discussion of the important time-dependent property of creep, it is found that particulate fillers, either through a hydrodynamic effect on viscosity, or an interaction effect more closely associated with elastic response, give rise to a reduction in creep. Up to the point where debonding is observed, the creep behaviour can be generally represented by the following equation:[12]

$$\epsilon(t) = \epsilon_1(t) . E_1/E \qquad (3.6)$$

where $\epsilon(t)$ and $\epsilon_1(t)$ are the extension of composite and matrix, respectively, and E and E_1 are the corresponding moduli. Beyond this point, the rate of extension increases until finally, fracture occurs. Another time-dependent effect observed with particulate-filled polymers arises from slow crack-growth for a material under stress, leading to static fatigue failure at lower stress levels than would be expected from normal tensile testing procedure. This deterioration in properties can be further exacerbated by the presence of moisture or other 'aggressive' environmental conditions.

Although debonding and crack formation lower the strengths in general of composites, in certain cases in which cross-linked low-energy brittle plastics such as polyesters and epoxies are involved, the actual fracture energy, as distinct from strength, may well be increased by the presence of filler, up to quite high concentrations of filler after which it decreases again. The effect is less for smaller particle sizes. It is believed that the presence of filler can actually impede crack growth for various reasons, including the possibility of imposing a greater tendency for plastic deformation in the matrix, the influence being stronger in the absence rather than the presence of bonding agent. At the high rates of strain associated with impact testing, the behaviour does not appear to be the same, the impact strength almost invariably decreasing with increase in concentration of hard particulate filler. The few cases in which rigid filler appears to produce a rise in impact strength seem to be associated with an ability to produce an increase in the cohesive strength of the matrix or where it can alter the distribution of stress to cover a larger area as sometimes can take place with very fine dispersions at low concentrations. The other and much more important technological case of improvement of impact behaviour is that brought about by the use of soft rubbery fillers. These have been used for over half a century to produce impact-resistance in otherwise brittle plastics, the understanding of their precise action lagging many years behind their commercial introduction. They have found particular use with polystyrene and polyvinyl chloride plastics, two of the early synthetic polymeric materials. The effect, directly on impact strength can be seen in Table 3.1

TABLE 3.1
IMPACT STRENGTHS OF POLYMERS AND COMPOSITES

Material	Impact strength (J/cm)	Material	Impact strength (J/cm)
Polystyrene	0·15	Polyester (woven roving)	2·5
High impact polystyrene	0·35–1·5	Polyester (chopped strand mat)	2·0
Polystyrene/polyphenylene oxide	1–2	Polyester (plain weave cloth)	2·5
Polystyrene (30 per cent glass fibre)	0·6	Phenolic resin	0·1–0·2
Polyvinyl chloride	0·35	Phenolic resin (paper)	0·2
Polyvinyl chloride (25 per cent ABS)	9	Phenolic resin (cotton)	0·3–0·5
Polyvinyl chloride (75 per cent ABS)	4	Polypropylene	0·5
ABS	1–5	Polypropylene	0·5
Polyethylene	1·25	(20 per cent short glass fibre)	
Polyethylene (40 per cent glass fibre)	0·7	Polypropylene	1·8
Acetal resin	0·7	(20 per cent long glass fibre)	
Acetal resin (40 per cent glass fibre)	0·35	Polyphenylene sulphide	0·15
Nylon 6	0·5	Polyphenylene sulphide	0·4
Nylon 6 (20 per cent glass fibre)	0·7	(40 per cent glass fibre)	
Nylon 6 (40 per cent glass fibre)	1·6	Epoxy resin	0·1–0·6
Polyester (cross-linked)	0·1	Epoxy resin (glass fibre)	5–15

Note: data taken from different sources. All sources do not agree, presumably due to differences in material, composite fabrication, and test method.

and indirectly through their influence on fracture energy, i.e. the energy under the stress–strain curve, in Fig. 3.4, which also shows the effect of temperature on impact strength. The mechanisms advanced for the increase of impact strength are numerous and for a particular case may be different, at least in degree, depending on the nature of the composite,

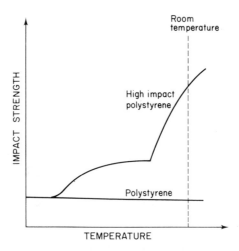

FIG. 3.4. Impact behaviour of brittle polymer and rubber blend.

temperature, and test conditions. One of the earliest suggestions was that the rubbery inclusion has the ability to absorb excess energy through a damping mechanism. However, this explanation does not account for the general whitening effect accompanying the usual impact response of composite polymers. Because of this, an alternative explanation invoking multiple crazing has been developed, in which it is proposed that the rubber initiates and controls craze growth.[13] The craze which is initiated at a point of maximum strain, typically near the equator of a particle, grows until it meets another particle, or, of course, when the stress concentration at the craze tip falls to zero. Thus, instead of producing large crazes leading to cracks, when a filler is present a large number of microcrazes are formed. Incidentally, crazes are also a feature of cyclic straining which ultimately lead to fatigue failure, but here the somewhat interesting phenomenon of craze healing or partial craze healing can occur during the compression part of the cycle, emphasising (in the same way that annealing by heat treatment can act) the craze-matter structure of the craze which distinguishes it from a true crack (Fig. 3.2). A third possibility which leads to enhanced impact resistance is that of shear yielding in toughened plastics, or a combination of crazing and yielding. The former is believed to be dominant in high-impact polystyrene, whilst the joint contribution is probably operative in ABS plastics. Other factors affecting impact behaviour of particulate composites of this type are particle size (larger particles being more efficient than smaller although there may well be an optimum size, which is presumably different for different systems), volume fraction with respect to maximum packing fraction, and adhesion between the two phases. Again this appears to have some optimum relationship as will be discussed later, in connection with the mechanical properties of fibre-filled polymers.

Before turning to these systems, however, some comment is appropriate for completion of this section, on the influence of the intermediate type of filler such as flakes, associated with some clay fillers, glass, mica, and some metals, and to particulate fillers which have irregular shape. In general, composites based on these tend to have higher moduli and higher strength for force directions parallel to any planes of orientation. They tend also to have lower impact strengths and yet may provide a facility for improved mechanical damping arising from layer slippage.

MECHANICAL PROPERTIES OF FIBRE-REINFORCED COMPOSITES

In many ways, fibre-reinforced polymer composites are by far the most important class of all composites, being specifically designed and used for applications which take advantage of the enhanced properties offered by properly fabricated materials, instead of, as for some particulate systems, being used to reduce cost. To appreciate their properties, particularly in relation to other materials of construction, attention is drawn to Fig. 3.5 and Table 3.2 in which emphasis is placed not only on the basic mechanical properties of modulus and strength, but also on specific modulus and specific strength, which, by any standards, are outstanding. A great deal of experimental and theoretical data is available now on the properties of composites resulting from micromechanical analysis carried out over the last twenty years or so, but before examining this, it is useful to consider the various factors which contribute to the mechanical behaviour of composites as a whole. Some of these have already been considered in previous chapters, but they are drawn together here to

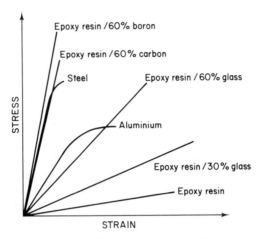

STRAIN

Epoxy resin / 60% boron

Epoxy resin / 60% carbon

Steel   Epoxy resin / 60% glass

Aluminium

Epoxy resin / 30% glass

Epoxy resin

FIG. 3.5. Tensile behaviour of fibre-reinforced plastics and other structural materials. (Reproduced with permission of Marcel Dekker, Inc.[14])

form a basis for appreciating their special effect on fibre-filled reinforced polymers.

There is first of all the matter of the chemical nature of both the filler and the matrix. Not only is there an infinitely variable capacity for change in the molecular structure of the latter, especially in some systems such as the unsaturated polyester or epoxy polymers, but even the filler may come in one of a number of different forms. This is illustrated by the variety in the range of glass fibres or even carbon fibres, just as for carbon blacks, which are on the market. Then the filler and resin may exhibit differences in physical properties within the class, since in the case of the matrix, there could be different degrees of crystallinity or differences in crystal and molecular orientation, whilst the filler might be, for example, either microcrystalline in nature or, in special cases, single crystal based. Generalising, both between and within the various classes of both components, one can say that, in the main, polymers contribute the properties of toughness, low density, low strength, low stiffness, high thermal expansivity, and low thermal stability in both a physical and chemical sense. On the other hand, fibrous fillers typically contribute high modulus, high strength, and often the disadvantage of brittleness. Together, the two components can be made to draw on each other's good properties, at the expense of their poor properties.

Next there are the variables of fibre length, length distribution, diameter, cross-sectional shape, surface irregularity within some average

TABLE 3.2
TENSILE MODULI AND STRENGTHS OF POLYMER COMPOSITES AND RELATED MATERIALS

Material	Modulus (GN/m^2)	Specific modulus (GNm/Mg)	Tensile strength (MN/m^2)	Specific tensile strength (MNm/Mg)
Steel	3000	385	210	27
Aluminium	150	55	69	26
Polyester (29 per cent chopped strand mat)	10·3	6·4	103	64
Polyester (20 per cent SMC)	13	9	75	50
Polyester (40 per cent glass fabric)	20	12	320	188
Epoxy (14 per cent S-glass fibre)	12	9	520	377
Epoxy (60 per cent S-glass fibre)	51	26	2000	1026
Epoxy (60 per cent boron fibre)	140	68	920	447
Epoxy (50 per cent carbon fibre)	190	129	590	401
Epoxy (40 per cent alumina whisker)	166	95	500	287
Nylon (40 per cent glass fibre)	11	8	200	142
Polystyrene (30 per cent glass fibre)	8	6	97	76
Polyester (80 per cent filament wound glass fibre)	55	28	1000	500

shape, and of course filler concentration, particularly with respect to maximum volume fraction theoretically possible within some given geometrical organisation, as outlined in Chapter 1. There it was shown that in addition to discontinuous and continuous single fibres, there are combinations of woven and non-woven mats and fabrics, each possessing potential for different fibre orientations, and in the latter different weaves. There is the further possibility of hybrid systems either with respect to mixtures of fibre types or fibres with particulate fillers. Then there is the way that separate layers may differ or be arranged in the preparation of a laminate. Many useful composites now finding application in automobile, aerospace, and marine engineering are fabricated in the form of laminates and the study of this aspect and the relative ways that different laminates may be brought together comes under the heading of what are called macromechanics. This is essentially an engineering topic and as such will not be discussed further. It is, however, well documented for the interested reader.[14]

Another factor which bears on polymer composite properties and one which is playing an increasing role in both experimental and theoretical branches of the subject, is that of adhesion between the polymer and filler and the way that this can be changed by the use of coupling and coating agents, affecting as they do, the way that stress is transferred from one component to the other. Closely associated with this, at least for practical purposes, is the mechanical adhesion which exists between the two, brought about by the greater relative volume changes which accompany the curing of a thermoset resin or the cooling of a thermoplastic polymer after moulding, as compared with the filler. For example, this may, in absolute shrinkage terms, be as high as 4 per cent for a cross-linked polyester or 1 per cent for polyethylene.

Other parameters which are of importance in affecting the mechanical properties are differences in sensitivity to environmental factors such as water, oxygen, ozone, and chemicals in general, as well as changes which can arise from changes of temperature. Again moulding conditions at the time of fabrication even for the same nominal moulding compound can affect behaviour. Since some moulding methods are more convenient with one system than another or perhaps more convenient for one fabricator than another, differences in properties, such as modulus and strength, can arise. To illustrate, if one method, for example, is prone to the possibility of void formation, then immediately there is an inherent element for the production of inferior composites. Finally, mention is made of the variability in response of composites arising from differences in the direction of the applied force under test or in use, or differences relative to some in-built anisotropy within the reinforced polymer.

In view of the large number of factors which can in principle affect the response of a composite to mechanical forces, it may come as a surprise to learn that the nature of many of these is fairly well known and in a great number of cases is susceptible to quantitative prediction. It is the purpose of what now follows to see to what extent this is so and to highlight some of the effects of adding a fibrous reinforcement in real systems.

Let us begin first of all with a consideration of stiffness, and in particular with Young's modulus. The modulus of the composite constructed from parallel and continuous arrangements can be calculated from those of the individual components. To do this (Fig. 3.6), assuming

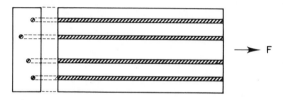

FIG. 3.6. Unidirectional continuous fibre-reinforced composite.

iso-strain conditions for matrix and fibre, that they both deform in an elastic manner and that Poisson's ratio is the same with no debonding taking place, on a work balance one can write:

$$Fl = F_f l + F_m l$$

where F, F_f, and F_m are the forces experienced by the composite, fibres, and matrix undergoing deformation l. In terms of stress, this can be rewritten:

$$\sigma A l = \sigma_f A_f l + \sigma_m A_m l$$

where σ, σ_f, and σ_m are the corresponding stress terms, and A, A_f, and A_m are the cross-sectional areas of the composite, fibres, and matrix. Rewriting and remembering that the strain is the same in all cases, we have:

$$E_c = E_f v_f + E_m v_m \qquad (3.7)$$

where E_c, E_f, and E_m are the moduli of the composite, fibre, and matrix. (NB. $v_f + v_m = 1$ where v_f and v_m are the volume fractions of fibre and matrix.) This is an example of the so-called Law, or Rule, of Mixtures. A limitation of the derivation was the assumption of identical Poisson's ratios for both components and that the materials are within their elastic limits. This should be true for elongations less than a few per cent and indeed, despite these limitations, the equation has been of value in predicting short strain effects on modulus.

Using essentially the same approach, but for iso-stress conditions, it is possible to calculate the corresponding relationship for the transverse modulus, E'_c,:

$$1/E'_c = 1/E_f v_f + 1/E_m v_m \qquad (3.8)$$

(The two approaches are essentially the series and parallel calculations commonly used in electrical and thermal conductivity problems.) In practice the transverse equation is much less applicable and alternative equations are preferred. One of the better known of these, and one which can by appropriate substitution be used for other modulus situations is that of Halpin and Tsai.[15]

Inspection of the simple equations will reveal that at the usual levels of fibre addition, the longitudinal modulus is dominated by the fibre modulus but the transverse modulus is more influenced by the matrix modulus. However, the transverse modulus is often much higher than predicted. For instance, where the longitudinal modulus of a fibre glass–epoxy resin composite is increased by a factor of 15, the same composite has a transverse modulus showing a reinforcement factor of as high as seven times that of the matrix, which means a more satisfactory performance than might otherwise have been anticipated.

To describe the intermediate case of a force direction which is not in either of the two directions indicated above, the Law of Mixtures can be written in the form:

$$E_c = \alpha E_f v_f + E_m v_m \qquad (3.9)$$

where α is an efficiency factor which can be evaluated for any angle of orientation. The influence of orientation can be seen in Fig. 3.8. The concept of an efficiency factor can also be used to embrace other possible variables having an influence on modulus, such as adhesion between polymer and filler, or mean angular distribution of the fibres, etc. Values of the efficiency factor, alternatively known as the Krenchel factor, have been tabulated and it has been shown, for example, that for the same polyester–glass fibre system, but with the glass fibre in the form of chopped strand mat, woven rovings, and unidirectional, for which α takes in turn, values of 0·3, 0·5, and 1·0, the modulus enhancement ratios are 2, 5, and 13. It should, however, be pointed out that in these the concentration of fibre is different in each case, being 0·2, 0·35, and 0·5, reflecting the different abilities of the various arrangements to take up resin, but even if the enhancement ratios are compared at the same concentration, the levels of the ratios are in the same sequence.

Turning to reinforcement by short, discontinuous, fibres particularly relevant to reinforced thermoplastic polymers as well as many thermosetting systems where the fibres, if anything, are usually somewhat longer, we can see that micromechanical analysis cannot be as straightforward since only the polymer matrix is continuous. Further, the process of fabrication, certainly with high-shear injection or extrusion moulding, can cause fibre breakdown so that the system is not what it might be expected to be on the basis of the starting materials. Because of the discontinuous nature of the reinforcement, the tensile stresses experienced by the fibres must derive from shearing forces transmitted by the resin. Clearly the greater the mean stress carried by the fibres the greater will be the modulus enhancement. It can be shown, as will be described below in a consideration of tensile strengths, that there is a critical fibre length, equal to the shortest length which will allow the stress in the fibre to reach the tensile fracture stress. The length depends upon the ratio of the modulus of the two phases, the strength of the interfacial bond, and the shear strength of the polymer. Typically for glass and carbon fibres it is quoted as being about 1–2 mm, but it will also depend upon aspect ratio. For this reason, asbestos with its small diameter is able to give a higher enhancement than glass, everything else being equal. However, in practice, the immediate advantage is offset by the fibre breakdown which readily occurs with asbestos on processing. The influence of aspect ratio is shown in Fig. 3.7, from which it appears that for maximum enhancement of modulus, and tensile strength, the aspect ratio should be above about 100. It has been shown by Cox[16] that the Law of Mixtures can be expressed in the following way to show the relationship between modulus and fibre length, separated from the orientation parameter:

$$E_c = \alpha_0 \alpha_1 E_f v_f + E_m v_m \qquad (3.10)$$

where α_0 and α_1 are orientation and length efficiency factors. A more complicated relationship can be derived to relate the latter to the possible distribution of lengths. The Cox–Krenchel equation shown

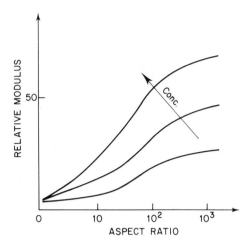

FIG. 3.7. Influence of aspect ratio on relative modulus (E_c/E_m).

above is reported to give good agreement with a large number of experimental results for short-fibre reinforced thermoplastics.[17]

So far attention has been directed to linear arrangements of fibres in the main, but as has been shown, the Young's modulus is parallel and at right angles to the fibre direction and those for other arrangements are all different. This is shown in Fig. 3.8. It must be emphasised, however,

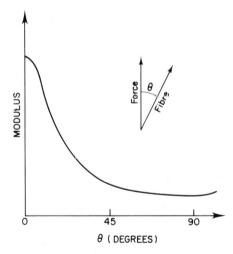

FIG. 3.8. Variation of modulus with angle between fibre and force direction in unidirectional continuous fibre-reinforced composite.

that for the case of woven reinforcement, the nature of the weave, quite apart from its secondary effect of limiting resin penetration, will also influence the composite modulus.

Considering now the other important tensile property, tensile strength, as we have seen in Fig. 3.5 the use of fibres not only increases modulus, but also strength, the greatest effect being obtained with unidirectionally arranged reinforcements as occurs in filament-wound and pultrusion-based composites. And again, as for modulus, theoretical analysis centred on micromechanical techniques has been carried out, but in general, the analysis is much more complicated. For this reason, the tensile strengths of unidirectionally reinforced polyester–glass fibre systems, for example, tend to be no higher than about 65 per cent of the values calculated on the basis of the Law of Mixtures, compared with almost complete agreement for modulus studies. The complications arise not only because of possible anisotropy and irregular disposition, such as in clustering, or by the usual environmental effects, but also because of the many different modes by which fracture can take place. These include the effect of structural variations in both matrix and fibre due to local weaknesses, voids, stress concentrations, etc. In addition, the nature and magnitude of interfacial adhesion, packing perfection, overlap of fibres, stress concentrations at the end of fibres, interference with crack propagation, and differences in plastic and elastic responses will have an influence.

On an energy balance consideration, as for the case of modulus, it is possible to frame a relationship for the tensile strength of a composite, in terms of the separate contributions of fibre and polymer. Once again assumptions as to perfect interfacial adhesion and to equality of Poisson's ratios are made, although it is realised that mismatch in Poisson's ratio can lead to triaxial stresses in the composite. Although fibre thickness does not come into the equation, it will have an effect, if only through the increased interfacial area, leading to an increase in strength. The equation, for a unidirectional orientation of fibre in the stretching direction is:

$$t_c = t_f v_f + t'_m v_m \qquad (3.11)$$

where t_c and t_f are the tensile strengths of the composite and fibre, and t'_m is the stress on the matrix at the breaking strain of the fibres, assuming that all fibres break at the same time. If the breaking strain of the matrix is less than that of the fibre, which would be exceptional for the usual composite systems of technological interest, the tensile strength of the composite would be:

$$t_c = t_f v_f \qquad (3.12)$$

Where the adhesion is less than perfect between the fibre and the matrix the Law of Mixtures can be written:

$$t_c = k v_f t_f + v_m t'_m \qquad (3.13)$$

where k is an adhesion factor, taking values between 1 and 0 corresponding to the limits of perfect and zero adhesion. This indicates the importance of matching fibre to resin, either directly or through the use of a coupling agent.

If the composite does not fracture when all the fibres are broken, as can occur at low concentrations, then the Law of Mixtures breaks down, the stress being associated entirely with the matrix. By equating the strength of the matrix to that of the composite, the critical volume fraction below which there is no reinforcing effect can be calculated.[18] It can also be shown that this limiting volume becomes smaller the greater the tensile strength of the fibre compared with that of the matrix. The composite tensile strength–concentration relationship is shown in Fig. 3.9. In many real systems, because of inherent variability of fibre strengths from one fibre to another, the tensile strength is less than that predicted from eqn. (3.11). Other deviations can arise from the fact that if a fibre breaks at its weakest point, it will leave two shorter fibres of

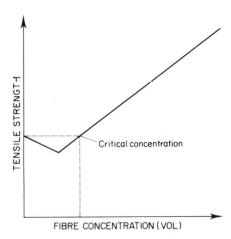

FIG. 3.9. Variation of tensile strength with concentration of fibre in unidirectional continuous fibre-reinforced composite.

strengths greater than the parent; also, there may be incidental changes in void formation, stress concentration, and changes in wetting accompanying tensile strain. Finally, it should be recognised that, perhaps because of a microcrack healing effect of the resin on the fibre surface, the measured strength of the free fibre may be less than that of the embedded fibre, complicating choice of t_f in use of the Law of Mixtures.

Attempts to calculate the transverse tensile strength have not been as successful as for transverse modulus, but a general equation which expresses tensile strength as a function of the angle between filament and applied strain directions has been derived by Tsai[19] and is shown graphically in Fig. 3.10(a). The transverse tensile strength is fairly closely

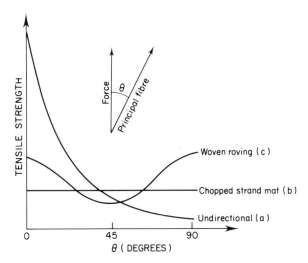

FIG. 3.10. Variation of tensile strength with stress direction for different composites.

related to matrix strength, being independent of fibre strength since ideally no fibres are broken in transverse fracture. It will be dependent also on the properties of the interface and, of course, on filler concentration. The variation of strength with orientation can formally be introduced in the Law of Mixtures in the usual way, as an efficiency factor. Its value will also be identified with other spatial organisation of the fibre, such as woven fabric or random orientation as in chopped strand mat. Again tensile strengths are lower than for the unidirectional systems but on the other hand, they tend to be less sensitive to variations in tensile strain directions as shown in Fig. 3.10(b) and (c), but the actual level of strength will also be limited by the amount of resin which can be accommodated by a particular geometrical arrangement. For example, although a unidirectional glass fibre–polyester laminate may have a tensile strength, say, of over $800 \, MN/m^2$, when tested in the orientation direction, compared with only half this value for a woven glass fabric, it should be appreciated that whereas the latter can only be incorporated at a 65 per cent concentration, the former will typically be at a 75 per cent concentration for sensible compositions.

In the case of discontinuous fibres, of growing importance in particular for fibre-reinforced high-performance engineering thermoplastics, a complication arises in mechanical analysis from the effects of the fibre ends. Irregularities in stress patterns develop, but although they constitute, in principle, stress concentration regions, these do not immediately affect tensile strengths of composites which are dependent on the transfer of stress to produce tensile stresses in the fibres by the action of shearing forces across the interface. Near the ends of the fibres, the shear forces are at a maximum, whereas the tensile forces at the ends are zero as indicated in Fig. 3.11. As the shear force strength decreases on going

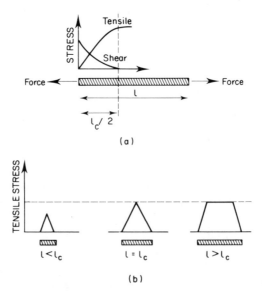

FIG. 3.11. Stress patterns for short fibre-reinforced composite. (a) Interface shear and fibre tensile stresses. (b) Variation in fibre tensile stress with fibre length.

along the fibre, the tensile stress in a particular fibre eventually reaches a maximum or steady value. Thus there is a certain length from each end before maximum stress is reached. The sum of the two form a critical fibre length, l_c. By considering the force required to overcome the shear force, the critical fibre length is calculated to be:

$$l_c = (t_f/2s_i)d \tag{3.14}$$

where s_i is the interfacial shear stress and d is the fibre diameter. It will be seen that the critical length will therefore depend upon interfacial bond strength, fibre strength, and thickness, and so will vary from system to system and with any coating which the fibre may have been given prior to compounding and moulding. If we assume that the tensile strength increases linearly from the ends of a fibre, which is approximately correct, then the average stress over the critical length is just half that of the fibre tensile strength and for fibres of length less than the critical length, the stress will be less. The average fibre stress at failure for fibres whose length, l, is larger than the critical length is given by:

$$\bar{t}_f = (l - l_c/2l)t_f \tag{3.15}$$

and the tensile strength of the composite using the Law of Mixtures equation will be:

$$t_c = t_f v_f(l - l_c/2l) + t'_m v_m \tag{3.16}$$

The fibre length, therefore must be of the order of some twenty times that of the critical length before, in terms of reasonable experimental error, the composite strength approaches that of the continuous fibre-reinforced composite. This is shown graphically in Fig. 3.12. In practice, because of the fibre ends, the ultimate tensile strength is not considered to exceed about 6/7 of that for a continuous fibre composite. Other factors which will influence the strength are variations in fibre orientation, fibre breakdown during processing, and dispersion of fibre, including overlap of fibre ends. Many of these will depend in turn on filler and matrix properties as well as on temperature and shear rates associated with fabrication. Attempts have been made to embrace these in the overall theoretical treatment, orientation being absorbed by a Krenchel factor, as before. However, local modes of failure in a real situation reduce the efficacy of these approaches. The theory may well be of more value in interpreting the actual level of strength after breakdown, rather than predicting the behaviour from first principles. A case in point is perhaps provided by a chopped strand mat-reinforced polyester resin

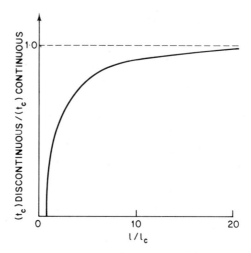

FIG. 3.12. Increase in composite tensile strength with increase of fibre length.

laminate which exhibits a linear stress–strain relationship up to a 'knee' above which some permanent debonding of transverse oriented fibre takes place, but the composite still presents a linear modulus, with elastic behaviour, although lower than before, until fracture (Fig. 3.13). Of all the factors which affect tensile strength, probably interfacial shear strength is the most important, so that any way in which this can be

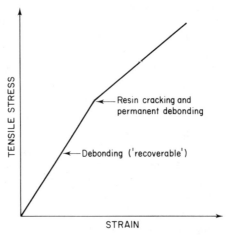

FIG. 3.13. Tensile behaviour of chopped strand mat composite.

improved, not only leads to higher tensile strengths, but by reducing the critical fibre length makes possible an improvement in the mechanical properties at the expense of overall fibre length which might otherwise have adverse effects on processing. It has been shown that for the extreme cases of resin behaving in an elastic manner and in a plastic way, the critical fibre length can vary from ten to a hundred times the fibre diameter.[20] To some extent interfacial shear strength will be influenced by the shrinkage effects on cooling from the mould or other shrinkage of the resin perhaps on curing, leading to a form of mechanical adhesion. It will be affected by fibre diameter through change in specific interfacial area. The shear strength is also affected by the presence of voids in the composite, decreasing linearly with void content, but the aspect which has created most interest in recent years has been the use of coupling agents on fibrous reinforcements, or modification of the fibre surface by, for example, chemical means as described in Chapter 2. To illustrate this, a 20 per cent glass fibre reinforced sample of polypropylene was found to more than double in tensile strength after the glass fibre had been coated with a coupling agent.

Because, as for other engineering materials, reinforced plastics are needed for use under compression as well as tension, there has been an interest in the compressive strength of composites. For continuous fibres aligned in one direction, it has been found that the strength under compression in this direction, is less than under tension. The basis of the failure when it occurs, has been attributed to shear yield in the resin, to debonding of the fibres, to voids, and to differential expansion effects arising from differences in Poisson's ratio. Failure can give rise to planar or spiral wave conformations in the fibres. For random distributions of short fibres, the compression strength and the tensile strength are about the same and follow a linear stress–strain relationship. Theories, such as that of Rosen,[21] which predicts a relationship of the form:

$$c_s = s_m/(l - v_f) \tag{3.17}$$

where c_s is the compressive strength and s_m is the shear strength of the matrix, whilst indicating both the role of the resin and the effect of filler concentration, tend to give low values for compression strength. It is possible, if special care is taken with the design of the experiment, that the appropriate form of the Law of Mixtures might apply.

Some indication has already been given as to the nature of failure in polymer composites and it is of interest to have a closer look at the actual fracture process. Whenever it occurs, whether by tension, compression, shear or impact, it is for one or more of the following reasons: cohesive failure of the matrix, cohesive failure of the filler, and adhesive failure at the interface or in the interfacial region. Fibre failure occurs when the stress in the direction of the fibre axis exceeds a certain value and usually leads to, or precedes, total failure in which resin cracking or interfacial debonding takes place. In some cases it is believed that the origin of fracture cracking is through microcracks produced by differential expansion effects attendant on cooling after moulding, or in subsequent thermal cycling. Some idea as to the mode of fracture can often be gained from the appearance of the broken specimen. For example, if fracture takes place at the interface, then fibres are pulled out and have little or no polymer adhering to them. On the other hand, for matrix failure, the fibres may again be pulled out, but this time there will be an indication of the presence of polymer on the fibres. In the cases where there is strong interfacial strength, fibre fracture may be observed, with very few ends being visible.

The toughening effect which fibres usually produce with resins, thereby possibly giving rise to a composite which is tougher than either component, has been attributed to the extra energy needed for pull-out, debonding or redistribution of stress, involving also the creation of new surfaces. Theories have been developed to put these, at least to some extent, on to a quantitative basis. In a particular situation, the mode of fracture may depend upon many different matters including fibre and matrix strength, load transfer efficiency, resistance to crack propagation, bond strength between fibre and matrix as well as such factors as the volume concentration of the fibre and its geometrical organisation. An example has already been given of this for a chopped strand mat reinforced polyester, where it was seen that debonding takes place at about 30 per cent of the ultimate tensile strength, but resin cracking did not appear until about 70 per cent of the breaking stress. In the fracture of polyesters and epoxy resins reinforced with woven glass fibres, debonding is observed to begin by debonding along those fibres at right angles to the strain direction and then at the crossover points of the weave. Finally, there is some straightening out of fibre crimp and debonding in a direction parallel to the strain, or fracture of the resin itself. In short fibre-reinforced thermoplastic polymer composites, it has been shown that for a wide range of polymers and fibres, fracture is mainly by fibres being pulled out of the matrix, whether brittle or ductile by normal standards. Cracks appear to form at the fibre ends and any misaligned fibres are pulled through the matrix, few fibres undergoing fracture themselves. It is interesting to note that ductile materials provide tough composites when the concentration is low, but at higher concentrations, toughness of the fibre itself becomes important. The process of fracture is shown in Fig. 3.14.

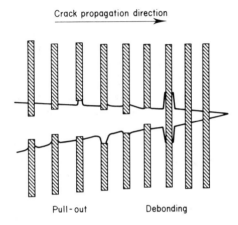

Crack propagation direction

Pull-out   Debonding

FIG. 3.14. Fracture process in fibre-reinforced plastic.

Toughening of the composite can be achieved in principle, in a number of ways, and where it may be lacking in one fibre–polymer system, it might be introduced at the expense of some other desirable property, but to the overall advantage of the composite, by the use of hybrid fibre systems. An interesting application of this concept is in the use of helically-wound glass and carbon fibres to simulate the structure of wood, to produce an epoxy resin composite of fracture energy, of the order of 10^5 J/m^2, comparable to that for ductile materials.[22] Thus one fibre, in this case carbon, can provide a high strength, whilst the other fibre, glass, besides being cheaper, can contribute an improved toughening to such a resin. Aromatic polyamide fibres are considered to give good balanced characteristics of both strength and toughness. Fibre length is found to be an important parameter in the toughening of polymers with discontinuous fibres, the best effect being developed at about the critical length. Another factor affecting toughness is interlaminar shear strength, with a maximum in overall toughness being reached with increasing shear strength. This indication of an optimum strength, has led to suggestions and some achievements for the design of tougher polymer systems. For example, if one lowers the adhesion between the fibre and resin matrix, a situation arises where appreciable work is expended in pulling broken or discontinuous fibres from the matrix. Application of a suitable coating on the fibre can modify interlaminar shear stress to improve toughness, although the same coating agent may not be the best, say for increase in modulus. Use of a 1 per cent silicone rubber coating on carbon fibre before compounding with a resin can more than double the work to fracture. This modification is not so dissimilar to the use of duplex fibres for toughening, in which one fibre is surrounded by another. Here, however, the toughening action comes from the energy dissipated in the frictional sliding of one fibre inside the other during deformation. In the same way that surface treatment of the fibre will, as mentioned above, improve composite toughness, so will attack at the interface reduce desirable mechanical properties. This environmental aspect, best illustrated by the influence of moisture on glass fibre composites, is discussed further in the next chapter.

A special branch of toughness, which has a particular relevance to polymeric engineering materials, concerns their behaviour under high strain rates, exemplified by impact testing. The use of both short and long fibres as reinforcing materials in an impact context is very important (see Table 3.1). In both cases except where the polymer itself, e.g. polyethylene, is inherently tough the impact resistance increases with increasing concentration, although eventually at high concentrations, the strength must fall again. Both the influence of concentration and the use of a coating agent are shown in Fig. 3.15. For oriented systems, fracture energy is greater for loading in the direction of the fibres, rather than at right angles, with the highest impact strengths being for short fibres of critical length, and for poor, as compared to strong, bonding. For transverse loading, however, good bonding is preferable. Clearly, because of this anisotropy in properties, random orientation may be an advantage for general applications. In the use of comparisons of this kind, it should be emphasised that data obtained from one somewhat restricted impact test procedure may not necessarily reflect behaviour in a practical impact situation for a real moulded article, so care must be exercised in the acceptance of apparent changes of properties.

Turning now to the subject of creep and fatigue in polymer composites, relevant to the limits of useful application, as for particulate-filled polymers, the amount of creep which takes place decreases with increase of fibrous filler. As before, the response can to a first approximation, be represented by eqn. (3.6), but this relationship must be used

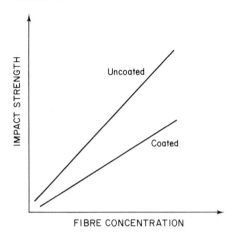

FIG. 3.15. Influence of fibre concentration and pretreatment on composite. (Note: pretreatment improves tensile strength.)

with some caution since very slight changes in test conditions in relation to use conditions can lead to substantial deviations. If at some stage debonding occurs, then the rate of creep increases as slippage at the fibre–matrix interface takes place. The effect of increasing filler concentration, as indicated by the above equation, will be a reduction in creep, the amount varying from one composite to another. In a typical reinforced plastic, the effect of filler is to reduce creep by a factor of two or more, but again this will vary with the particular system. For example, for a sample of polyethylene terephthalate containing 30 per cent glass fibre, the strain after a given time was found to have fallen by a third of the value for the corresponding unfilled polymer, whereas for nylon 66, a polymer with a greater intrinsic creep tendency, the fall for the slightly higher concentration of 35 per cent, was only one eighth that of the base polymer.[23]

We have seen that static loading not only gives rise to creep but also to fatigue failure, which may masquerade as material creep. Fatigue is a subject much studied in all structural materials over the last thirty years or so in view of its great importance. The origin of failure tends to be quite different in different materials, metals, ceramics or plastics and can vary considerably in the last class. The fatigue failure of polymer composites (most work appears to have been carried out on laminates) can be studied in different modes, but one of the most common experimental approaches has been to apply a cyclic stress and then record the time to failure in terms of the number of cycles, as illustrated by S–N curves (Fig. 3.16). Despite the large amount of work done, much of the information generated is used in an empirical way and does not always give a satisfactory indication of failure mechanisms. One of the problems is in deciding what are the criteria for failure, the significance being of unequal importance in different situations. Then there are different types of failure as we have already seen, and indeed, it is possible that a certain amount of fracture may well result in increased toughness in some cases.

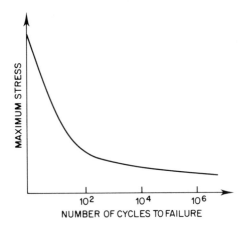

FIG. 3.16. Typical S–N curve for a plastic.

From studies which have been carried out it would seem, however, that the two factors contributing most to failure are matrix fracture and weak bonding between matrix and fibre, and therefore, a considerable amount of effort has gone into the improvement of bonding, or to prevent access of debonding agents, such as water, to the interface. Since fatigue failure is a progressive matter, attempts have been made to put the amount of damage on to a quantitative basis. An equation which does this is:[24]

$$D = \sum \left(A\frac{n}{N} + B\frac{n^2}{N} \right) \tag{3.18}$$

where D refers to the damage, A and B are constants, and n is the number of cycles which the composite has experienced, whilst N is the number normally expected to produce failure at a given stress. Inherent in this is the acceptance that fatigue is an accumulation of damage and data have been published on this topic for a wide range of systems of both continuous and discontinuous fibres in thermoplastic and thermoset polymers. From this, it has been shown, for example, that although debonding frequently, if not always, takes place within a few cycles, cracking might not occur, at least in an observable manner, for up to perhaps, a thousand cycles, and ultimate failure possibly not until ten thousand cycles. Smith and Owen[25] have shown that S–N curves pertinent to each sequential mode of 'failure' can be constructed (Fig. 3.17). Considering some of the factors which bear upon fatigue, stress

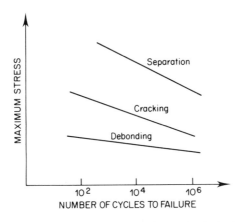

FIG. 3.17. S–N curves for chopped strand mat/polyester.

level is of importance as for unfilled systems, with higher stress levels tending towards a shorter life. The presence of notches and holes, of special relevance to user application, with intrinsic stress concentration implications, also limits life. The magnitude of the strain in cyclic stressing will also have a bearing on fatigue life. Increasing concentration of filler moves the S–N curve up the stress axis as shown by carbon fibre-reinforced nylon 66 composites, and weight for weight, carbon is more effective in this than glass.[26] Increase of temperature generally produces a deterioration in fatigue life and since cyclic stressing can give rise to dissipation of energy, the breakdown through temperature increase may not necessarily be because of any change in ambient temperature. Environmental factors of great and critical importance are water, chemicals, and radiation, especially UV radiation, all typically producing premature failure. Frequency of test cycle also has an influence, the effect being greater in many cases, the lower the frequency. Fibre stiffness can change the shape of the S–N curve. Fibres like graphite and boron which are both stiff absorb the greater proportion of applied stress, and produce more horizontal curves than does glass. In the better fatigue-resistant behaviour of these fibres, it is possible that the greater thermal conductivity may also play a part by dissipating heat. Aspect ratio increases generally produce longer fatigue lives, eventually levelling out at some particular value of aspect ratio, e.g. about 500 for a boron–epoxy resin composite. Fabrication method, involving differential shrinkage of resin and fibre will, as for other properties, affect fatigue, and the actual nature of the polymer will have a further influence in that an essentially brittle polymer will be more prone to failure by cracking than will be a ductile polymer where debonding may be of greater importance.

In this chapter an attempt has been made to give a broad review of the mechanical properties of polymer composites, and of necessity certain areas have had to be excluded. These, which include for example, concepts of macrocomposites and, as mentioned earlier aspects of the macromechanical approach to the properties of laminates, the continuing interest in property-forecasting from non-destructive testing, etc., are well documented in the scientific and engineering literature on composites, to which separate reference should be made.

REFERENCES

1. CARSWELL, T. S. and NASON, H. K., *Modern Plastics*, **21**, 121 (1944).
2. GRIFFITH, A. A., *Phil. Trans. Roy. Soc.*, **A221**, 163 (1920).
3. BUCKLER, E. J., *Plastics & Rubber Int.*, **4**, 255 (1979).
4. EINSTEIN, A., *Ann. Physik*, **19**, 289 (1906); **34**, 591 (1914).
5. MOONEY, M., *J. Colloid Sci.*, **6**, 162 (1951).
6. See MASCIA, L., *The Role of Additives in Plastics*, p. 87. Edward Arnold, London (1974).
7. KERNER, E. H., *Proc. Phys. Soc.*, **B69**, 808 (1956).
8. LEWIS, T. B. and NIELSEN, L. E., *J. Appl. Polymer Sci.*, **14**, 1449 (1970).
9. HASHIN, Z. and SHTRIKMAN, S., *J. Mech. Phys. Solids*, **11**, 127 (1963).
10. Ref. 14, Chapter 1.
11. Ref. 2, Chapter 2, p. 113.
12. NIELSEN, L. E., *Trans. Soc. Rheol.*, **13**, 141 (1969).
13. BUCKNALL, C. B., *Toughened Plastics*, Applied Science Publishers, London (1977).
14. See, e.g., MCCULLOUGH, R. L., *Concepts of Fibre-Resin Composites*, p. 16, Marcel Dekker, New York (1971).
15. HALPIN, J. C. and TSAI, S. W., *Primer on Composite Materials: Analysis*, Technomic Publ. Co. (1969).

16. COX, H. L., *Brit. J. of Applied Phys.*, **3**, 72 (1952).
17. DARLINGTON, M. W. and SMITH, G. P., *Fibre Reinforced Materials*, p. 63, Instn. of Civ. Eng., London (1977).
18. Ref. 6, p. 72.
19. See also JOHNSON, A. F., *Engineering Design Properties of GRP*, p. 105, British Plastics Federation, London (1979).
20. TSAI, S. W., Monsanto/Washington Univ. ONR/ARPA Assoc. Doc. HPC 68 (1968).
21. ROSEN, B. W., *Fiber Composite Materials*, Chapter 3, Am. Soc. for Metals, (1965).
22. GORDON, J. E. and JEROMIMIDIS, G., *Phil. Trans. Roy. Soc.*, **A294**, 545 (1980).
23. TITOV, W. V. and LANHAM, B. J., *Reinforced Thermoplastics*, Applied Science Publishers, London (1975).
24. OWEN, M. J. and HOWE, R. J., *J. Phys.* (D). *Appl. Phys.*, **5**, 1637 (1972).
25. SMITH, T. R. and OWEN, M. J., British Plastics Federation 6th RP Conf., Brighton (1968).
26. THERBERGE, J. ARKLES, P. and ROBINSON, R., *Ind. Eng. Chem. Rod. Res. Div.*, **15**, 100 (1976).

Chapter 4

PHYSICAL AND CHEMICAL PROPERTIES OF POLYMER COMPOSITES

INTRODUCTION

Mechanical properties constitute the main area of interest as far as composite polymeric systems are concerned, at least in terms of the amount of effort which has gone not only into designing materials for specific applications, but also in trying to interpret the behaviour from a knowledge of the properties of the individual components. Nevertheless there has been a great deal of interest in the other properties of composites. The reason for this is that in many cases fillers are incorporated into polymers for the purpose of modifying and improving both physical and chemical properties as well as upgrading the mechanical properties, or for reducing the cost of the final material. For example, electrically conducting fillers such as carbon are frequently added to what would otherwise be highly insulating substances having a tendency to accumulate static electrical charge, sometimes with disastrous consequences. A particular use of a carbon-filled plastic for this purpose is in polyvinyl chloride belting in conveyors for coal mines. Pigments are used to improve appearance and to reflect sunlight which can give rise to chemical degradation, with ultimate mechanical failure of a polymer. However, the use of a material for one reason can frequently produce a deterioration in some incidental property. So the nature of the filler, its concentration, and its interaction with the polymer may necessitate some compromise in order that immediate advantages in one direction are not offset by disadvantages in another. This chapter is devoted to a range of physical and chemical properties which are affected by the presence of fillers, both directly and indirectly. Amongst the former, we shall include the thermal, electrical, optical, magnetic, acoustic, friction, and wear properties, whilst under the latter mention will be made of chemical attack, including the action of water and oxygen, on composites and chemical changes which might be brought about by, for example, electric and mechanical stress.

In general, although the practical consequences of adding fillers to polymers in a physical context are only too well known, either from previous related experience or by intuition, the theories which try to predict this effect, certainly in quantitative terms, are much less advanced than they are for the mechanical properties of composites. Thus, at the present time, a great deal of empiricism colours the use of fillers in this area. As a first approximation, use is often made of the Law of Mixtures written in the general form:

$$P_c = \sum P_i v_i \qquad (4.1)$$

where P_c and P_i are some property relating to the composite and its components respectively, and v_i is the volume fraction of the latter. Not only is there the dilemma exemplified by whether one should be considering perhaps conductivity rather than resistivity, and so should be invoking use of the reciprocal equation:

$$1/P_c = \sum v_i/P_i \qquad (4.2)$$

but there is also the problem, as for mechanical properties, of the influence of associated factors. These include filler shape, size, size distribution, orientation, packing geometry, specific interaction between matrix and filler, and transfer effects across the interface. For this reason we should not expect a unique equation to describe the physical and chemical properties of composites in every aspect, and at the best can only hope that the various effects can be expressed by one or more adjustable parameters to the above or some other general equation.

THERMAL PROPERTIES

A great number of what might be termed thermal properties have an important bearing on the role and application of polymers and their composites. Not all of them will be discussed here, since in some cases little information on them is available outside, say, industrial laboratories. In addition, where data are available they may be of little use for analytical purposes, because of lack of precise information on the nature of the components, or absence of detail relating to fabrication techniques used in the preparation of a test sample. Perhaps to highlight a difficulty in the latter respect we could consider the measurement of the electrical conductivity of polymers. It is not uncommon practice to quote resistivity from a measure of current following the application of a potential across a specimen, and yet considerable evidence is available to show that it can take hours, even days, for electrical current equilibrium to be established. Thus, for any true structure–property relationship to be determined, a great deal of caution may be required in applying published data. In the same subject area and in the measurement of thermal conductivity of polymers, details of the nature of the guard plate used, if indeed a plate is used at all, are not always given. This also can lead to uncertainty in the validity of experimental results. Returning to a consideration of the thermal properties of polymer composites, the properties which will be discussed include thermal expansion, thermal conductivity, specific heat, softening including melting, thermal degradation and related phenomena.

Beginning with *thermal expansion*, the importance of this is that polymers and their composites often undergo comparatively large volume changes with respect to temperature during fabrication, whether through cross-linking or curing reactions in the case of thermosets or by the need for heating for the moulding of thermoplastics. In addition, polymeric materials are required to be suitable for application often over a wide range of temperatures, perhaps a few hundred degrees, both above and below room temperature. Since polymers as a whole have much larger coefficients of thermal expansion than many other structural materials, there can be quite a mis-match in expansion between polymer systems and the materials with which they might be in contact. For example, an elastomer with a coefficient of expansion approximately a twentieth that for steel can have a volume change on moulding, such that the final moulded article will be 2 per cent smaller in volume than the mould in which it was shaped. On the other hand, an epoxy resin containing about 60 per cent alumina filler has an overall thermal expansion similar to that for aluminium, so that in this case any differential shrinkage of the polymer composite in contact with aluminium would not be serious. However, at the local level there is the related problem of possible mis-match between the coefficients of the matrix and filler themselves. Because of this, a surprising amount of attention has been given in scientific laboratories to the question of the thermal expansion of composites contrary to what might at first sight be expected. One reason for this is that, as has already been indicated, because of differential shrinkage of filler and polymer, cooling of a composite in which there may or may not be good adhesion between the phases, or heating one in which there is good adhesion, can produce significant stress changes which in turn may have a profound effect on mechanical properties. Sometimes these stresses will lead to microcracks at the interface or nearby, and thus prepare the composite for premature failure. Annealing might well ameliorate the problem in some cases at one temperature, but only to transfer it to another. For our purposes we shall need to refer to linear coefficients of expansion (l) or bulk (α), the relationship between the two being for an isotropic material:

$$l = \alpha/3 \qquad (4.3)$$

and for systems of three-dimensional order:

$$\alpha = l_1 + l_2 + l_3 \qquad (4.4)$$

or for the more general case of unidirectional order in polymers:

$$\alpha = \frac{l_\parallel + 2l_\perp}{3} \qquad (4.5)$$

where l_1, l_2, l_3 refer to the three spatial directions, and $l_\parallel$ and $l_\perp$ refer to longitudinal and transverse directions.

To provide some idea of the magnitude of these values, data for various materials and composites are given in Table 4.1. Before considering thermal expansion in more detail, it might be useful to draw attention to the various factors which can affect expansion and in so doing can limit any theoretical approach. In addition to the intrinsic material contribution, there are the possibility of microcracks, the nature of the interfacial interaction, macrovoids in the case of expanded polymers (see Chapter 7), filler shape and orientation, temperature in relation to any thermal transitions in the polymer, and any molecular or crystal orientation within the polymer itself. Where there is a transition there will also be a change in the coefficient of expansion, and where there is some orientation, whether in the polymer or with respect to the filler, the coefficients will be different in the two orthogonal directions. A practical consequence of this in moulding is that warping may occur on cooling.

Considering first of all the thermal expansion behaviour of particulate-filled polymers, a number of different equations relating to the properties and concentration of the individual components have been published,

TABLE 4.1
LINEAR THERMAL EXPANSION FOR VARIOUS MATERIALS

Material	$\alpha \times 10^5/K$	Material	$\alpha \times 10^5/K$
Fused silica	0·05	Polycarbonate	6–7
A glass	1·0	Polyester	10
Iron	1·2	Polyester (CSM glass fibre)	1·8–3·5
Aluminium	2·5	Polyester (SMC glass fibre)	2·5
Copper	1·5	Polyester (WR glass fibre)	1·0–1·6
Epoxy resin	5–10	Polyester (UD glass fibre)	0·5–1·5
Polystyrene	8	Polyester (DMC glass fibre)	2·3–3·4
Polypropylene	10	Polyester (glass fabric)	1·1–1·6
Polypropylene (30% glass fibre)	4	Polyester (40% carbon fibre)	1·4
Polytetrafluoroethylene	14	Melamine aminoplast	—
Polyethylene (high density)	12	Melamine aminoplast (60% α-cellulose)	2–5
Polyethylene (40% glass fibre)	5	Phenoplast	—
Rubber	20–25	Phenoplast (40% α-cellulose)	2·9
Nylon 6,6	—		
Nylon 6,6 (40% carbon fibre)	1·4		
Nylon 6,6 (35% glass fibre)	2·4		

their usefulness often being limited to a single or restricted range of situations. Where there is no adhesion or mechanical interaction across the interface, the coefficient of expansion will be a constant independent of concentration, the matrix expanding away from the filler. Only when the temperature falls below that of the initial temperature will the deviations arise. This situation, as far as is known, is not found in normal composites. Where the matrix is relatively mobile, strictly as in a liquid but approximately as in an elastomer, the coefficient will fall with increasing concentration as predicted by the Law of Mixtures:

$$\alpha_c = \alpha_p v_p + \alpha_m v_m \qquad (4.6)$$

where α_c, α_p, and α_m are the coefficients of thermal expansion of the composite, particulate filler, and matrix. This equation is applicable to some natural rubber systems. In most composites, the determined value falls below that predicted from the Law of Mixtures. For these a number of equations have been proposed,[1] the best known being one due to Kerner which may be considered a deviation from the above:

$$\alpha_c = \alpha_p v_p + \alpha_m v_m - (\alpha_m - \alpha_p) v_m v_p \frac{(1/B_m - 1/B_p)}{v_m/B_p + v_p/B_m + 3/4G_m} \qquad (4.7)$$

where B and G refer to bulk and shear modulus.

The more the particles deviate from spheres, the greater the divergence is there from this equation. In such cases use has been made of an equation due to Thomas:

$$\log \alpha_c = v_p \log \alpha_p + v_m \log \alpha_m \qquad (4.8)$$

this relationship being completely empirical, or to an equation derived by Turner:

$$\alpha_c = \frac{\alpha_p v_p B_p + \alpha_m v_m B_m}{v_p B_p + v_m B_m} \qquad (4.9)$$

The shapes of the curves relating to these equations vary considerably as has been shown by Holliday and Robinson who have combined the above equations and introduced an adjustable parameter to provide an equation which is claimed to be of value to engineers, when the value of the parameter has been determined from a single point measurement. Concerning deviations from one curve to another, they conclude that particulate fillers have a minor influence compared with fibres and fabrics.

In relation to fibres, equations have been reported by Greszczuk[2] and Schapery,[3] that due to the latter taking the form:

$$\alpha_c'' = \frac{E_m \alpha_m v_m + E_f'' \alpha_f'' v_f}{E_m v_m + E_f'' v_f} \tag{4.10}$$

where α_c'' is the coefficient of expansion parallel to the fibre orientation direction and E_f'' and α_f'' are the longitudinal modulus and expansion coefficient of the fibre. There is a corresponding equation for the transverse direction. It will be noticed that the form of the equation is similar to that of Turner. The Schapery equation may be modified by an 'efficiency' multiplier term to precede the right-hand terms in the numerator and denominator, to include the effect of fibre length. As yet, no satisfactory interpretations of more complicated and agglomerated situations are available, although data on these have been published. For example, compared with a coefficient of linear expansion for a polyester resin equal to about $10^{-4}/°C$, typical chopped strand systems have values approximately one-quarter of this, woven roving systems, one-tenth, and unidirectional glass fibre composites about one-twentieth the value of the unfilled resin. To conclude, although most thermal expansion values are positive, reducing with increase of filler concentration, it has been reported that a phenolic–asbestos laminate may have a negative value over a narrow range near to room temperature.[4]

The second aspect of thermal properties which has received some interest as far as composites are concerned is that of *thermal conductivity*. The main activity here has been in connection with the use of polymers for insulation, particularly expanded polymers. But there is a growing interest in the effect of solid fillers on the thermal conductivity of polymers, both from the point of view of optimising heat input into the processing of thermoplastic and thermosetting polymers as well as increasing the coefficient of thermal conductivity for applicational purposes. This is attracting special attention at the present time with regard to rubber processing.[5] Polymers as a class have low thermal conductivity compared with, for example, metals and many inorganic materials (Table 4.2), but through the incorporation of, say, metal or carbon filaments, quite a substantial increase in conductivity can be obtained. To illustrate, whereas 30 per cent of glass fibre will raise the thermal conductivity of nylon 6 by about a factor of $\times 2.5$, the same amount of carbon fibre is about 10–15 times more effective.[6]

A number of theoretical treatments have been made in an attempt to relate the thermal conductivity of a composite to that of the individual

TABLE 4.2
THERMAL CONDUCTIVITIES OF VARIOUS MATERIALS

Material	Thermal conductivity (W/mK)	Material	Thermal conductivity (W/m K)
Copper	400	Epoxy resin	0·23
Aluminium	230	Epoxy resin (UD carbon fibre 60%)	61
Glass	ca. 0.9	Epoxy resin (glass fabric 43%)	0·38
Polyethylene (HD)	0·63	Epoxy resin (200% by weight Aluminium)	0·92
Polyethylene (LD)	0·33	Epoxy resin (powder)	
Nylon 6	0·31	Polyester	0·2
Rubber	0·18	Polyester (CSM)	0·16–0·26
Polystyrene	0·15	Polyester (SMC)	0·16–0·26
Foamed polymers	0·05	Polyester (WR)	0·2–0·3
Phenoplast (54% glass fibre)	0·60	Polyester (UD)	0·16–0·35

components and the concentration of the filler. The first was an equation derived by Maxwell[7] and relates to dilute systems of spherical particles. This equation, which incidentally has been used to analyse thermal conductivity data for polyethylene in terms of the separate contributions of amorphous and crystalline phases, has been extended by other workers. Different approaches to the problem have also been made and a number of other relationships have been reported, notably one by Nielsen[8] which takes the general form of the modified Kerner equation for stiffness reinforcement (see Chapter 3) in which modulus is replaced by the coefficient of thermal conductivity. Sundstrom and Chen,[9] comparing a number of equations, have shown that for glass-filled polystyrene and polyethylene an equation due to Cheng and Vachou[10] gave the best agreement with experimental results. Pointing out that thermal conductivities of composite polymers tend to have values which lie between the two limits of the Law of Mixtures, Ziebland[11] has suggested the use of the arbitrary relationship:

$$\log k_c = v_p \log k_p' + v_m \log k_m \tag{4.11}$$

where k_c and k_m are the thermal conductivities of the composite and matrix and k_p' is the hypothetical thermal conductivity of the particulate filler. The value of this can be determined from a one-point solution for a single reading on a composite of known composition. Once evaluated, the figure can be used for calculations involving other concentrations. Good agreement is reported with experimental results and k_m' assumes not unreasonable values. The similarity of this equation with that of Thomas reported above, will be noted.

In connection with the influence of other variables, which incidentally have received less theoretical attention, considering first the effect of particle size, it would appear that this has little bearing on thermal conductivity provided the thermal conductivities of the two components are not greatly different. In the case of a copper-filled epoxy resin, the same is true around ambient temperature but at low temperatures the larger contact resistance of the smaller particles reduces the conductivity. The surface is also important in, for example, copper and aluminium metal powders, where oxide-free and elongated particles have higher conductivities. For fibrous composites, in aligned carbon fibre–epoxy resin systems, it has been shown that the thermal conductivity in the parallel direction is of the order of fifty times that in the lateral direction[12] and that continuous fibres give rise to higher conductivities than discontinuous, the latter having a smaller longitudinal to transverse ratio also (Fig. 4.1). Comparing glass fibre to carbon fibre, the conductivity is less as is the longitudinal:transverse ratio again and glass fibre has a lower conductivity also than a metal fibre such as aluminium.[13] It has been reported that the thermal conductivity of phenolic–asbestos laminates increases with density as, of course would be expected

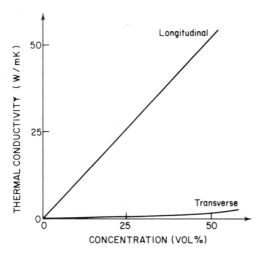

FIG. 4.1. Thermal conductivity of unidirectional carbon fibre/epoxy resin composite.

on the basis of the increased asbestos content assuming no concomitant significant void formation in the composite.[4] In the case of woven and cross-ply reinforcements, the equation due to Ziebland (eqn (4.11)) is found to give good agreement.

Not a great deal appears to have been published on the *specific heats* of polymer composites although data on this will be of importance to engineers concerned with transient flow in composites in view of the relationship:

$$\text{thermal diffusivity} = kd/C \qquad (4.12)$$

where d and C are the density and specific heat of a material. For a compact polymer composite the specific heat can be calculated from:

$$C_c = v_p C_p d_p + v_m C_m d_m \qquad (4.13)$$

In addition to temperatures associated with polymer breakdown, to be discussed later, the two most important *thermal transitions* relating to polymer systems are the melting point and the glass-transition temperature. Related to and dependent upon these are the technically important material characterisation temperatures of heat distortion temperature and softening point, which are temperatures at which the polymer undergoes some arbitrary deformation for a given load under specified experimental conditions. In practice, a softening point such as the Vicat softening point reflects the influence of matrix more than does heat distortion temperature where composites are concerned, and so changes tend to be smaller.

Melting point, say for crystalline polymers such as nylon, polypropylene, and acetal polymers, is largely unaffected by the presence of conventional fillers, as might be expected since the crystalline material will normally remain the same, unless for some reason the filler can affect crystal habit. However, if the filler can act as a nucleating agent for crystallisation (or in rare cases, an anti-nucleating agent) then by affecting crystallite size or perfection, it can influence not only the melting behaviour, but by any further effect on spherulite size, also those mechanical properties closely dependent on morphology. In practice, since nucleating efficacy reaches a maximum at concentrations not much above 1 per cent, and since most polymers contain other additives to about this level anyhow, the further effect of filler will probably be subsumed within the additive influence as a whole.

Generally speaking, the glass-transition temperature is less sensitive to the presence of filler than is heat distortion temperature, usually rising as measured by, say, differential thermal analysis, but after a few degrees becoming constant more or less independent of concentration, whereas heat distortion temperature may increase by many tens of degrees. A possible origin for this different response can perhaps be detected from a consideration of the mechanical loss curves of filled polymers. In these it is usually observed that although the position of the peak, expressed as a function of temperature, only changes slightly, an asymmetry develops, as mentioned previously, such that the mechanical loss at temperatures above the peak, identified with the glass-transition temperature, increases at the expense of the loss at temperatures below the peak. If the increase is identified as due to interaction between filler and polymer molecules (particle–particle friction could also be a factor), then it would appear that there is a fall in the mobility of the slowly moving segments of the polymer, possibly identifiable with an adsorbed layer on the filler surface, whilst the rest of the polymer is unaffected. Thus the glass-transition temperature is hardly changed, but the overall mobility is reduced. This means that the normal mechanical response to deformation as in the heat distortion test will be delayed, giving rise therefore, to a much enhanced heat distortion temperature, or, alternatively, softening point. Modification of the interaction by use of a coating agent could well change the actual behaviour. For a crystalline polymer, in which there is already some restriction of chain movement, the filler produces further restrictions.

It has already been mentioned that the increase in heat distortion temperature and softening point through the incorporation of fillers into polymers can be quite substantial, and since it is not uncommon to use these temperatures in formulating design limits in the use of plastics, then these changes have an important applicational consequence. A curve showing the effect of asbestos filler on the Vicat softening point for polystyrene is shown in Fig. 4.2.[14] For polyethylene, whilst 30 per cent

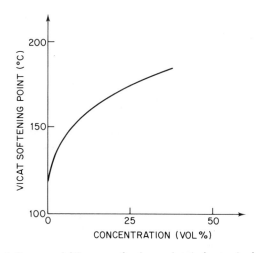

FIG. 4.2. Influence of filler on softening point (asbestos/polystyrene).

calcium carbonate is reported to raise the heat distortion temperature by about six degrees, the same quantity of glass fibre will increase the temperature by about eighty degrees. Similarly 30 per cent of glass or carbon fibre will raise the heat distortion temperature of polyethylene terephthalate from 70°C to 220°C and 40 per cent of carbon fibre will increase the corresponding temperature of nylon 6 from 70°C to 260°C, i.e. to about the degradation temperature.[15] Concerning other variables, for nylon composites, a threefold increase in glass fibre diameter can produce an increase in the heat distortion temperature of about 5°C with a corresponding decrease for polypropylene. Similarly, an increase in fibre length from 5 to 40×10^{-2} cm has no apparent effect for polypropylene, but causes a fall in heat distortion temperature of about 27°C for nylon, a surprising result in the light of the above results for polyethylene with particulate and fibrous filler and for the expected behaviour on the basis of mechanical properties. No doubt current work on Reinforced Reaction Injection Moulding (RRIM), to be discussed later, will throw some light on these anomalies. Finally in connection with more complex systems, chopped strand mat and sheet moulding compound grades of heat resistant polyester formulations, will increase the heat distortion temperature from 140°C to about 175°C, whilst woven roving and unidirectional composites are reported to increase the heat distortion temperature to about 250°C and 260°C respectively, these no doubt being upper practical limits in the light of impending polymer breakdown. It does, however, highlight the possibilities which might be attained by the use of high-temperature-resistant polymers.

Since this next transition, i.e. polymer degradation, is one of chemical breakdown, further discussion on it will be left to later in this chapter, but before leaving the subject of thermal properties, it is appropriate to conclude with some mention of another area of the thermal properties of polymer composites which has been of growing importance in the last few years, that is the use of composites at cryogenic temperatures, the interest being in connection with the use of liquefied gases and in spaceflight. Applications of composites are to be found as insulators and supports for superconducting magnets, as non-metallic Dewar vessels of complicated shape and as high-pressure vessels. Of the various polymers investigated, epoxy resins reinforced with carbon or glass fibre appear to have created most interest, carbon fibre–epoxy resin systems of suitable design having extremely small changes in contraction as the temperature is reduced from room temperature to cryogenic temperatures, but with a fall in fracture strain from 5 per cent to 2 per cent at 77 K and both glass fibre and carbon fibre composites having an increase in interlaminar shear stress. It has been found that with glass fibres, the cracking due to thermal shock does not prove a problem, providing the amount of epoxy resin used is small.

ELECTRICAL PROPERTIES

Although important in their own right as engineering materials, one of the main reasons for the development and growth of synthetic polymers in the first place was their outstanding electrical properties, especially their high resistivity. Because of this, many plastics can tolerate appreciable reinforcement or incorporation of filler to improve other properties without seriously affecting the useful electrical properties. Thus we find laminated plastics used in a range of insulating applications in domestic and industrial fixtures, transformer coil forms, terminal blocks, etc. The resistivity can in fact cause problems, as already mentioned, particularly for hydrophobic polymers in that it leads to a build-up of electrical charge. This can prove hazardous in certain powder industries and the transport of inflammable fuels, and it can prove troublesome in the textile spinning industry. Because of this, in addition to the use of certain antistatic agents, conducting fillers are sometimes used to reduce static electrification. This has assisted research into electrically conducting composites and recently they have been used as shields for electronic devices, against electromagnetic interference. The other two electrical properties which are of importance, in addition to conductivity and triboelectricity, are dielectric strength and permittivity.

Considering first of all the subject of *conductivity*, or its reciprocal, resistivity, there is an immediate difficulty in analysing data because of the very low level of conductivity and, as already outlined, the problem of reliability of results. Because of this, there is uncertainty as to the basic mechanism of conduction in composites, with small structural changes or apparently slight environmental effects having a dramatic influence on conduction. Thus, for example, an impurity level of a few parts per billion of some conducting species can theoretically raise the electrical conductivity by many orders of magnitude. Despite these problems, a number of equations relevant to specific systems have been reported, such as one due to Scarisbrick[16] which gave good agreement for a polyethylene–carbon black composite. Others for carbon in natural rubber, a most important applicational area, have been reported by Bulgin and by Studebaker.[17] The Law of Mixtures has been found to apply to aligned carbon fibre–epoxy systems. Processing conditions related to the amount of shear, orientation, and whether or not a dry or solution compounding process was used in the formulation, have also a bearing on properties.

Summarising some of the factors which affect the electrical conductivity of polymer composites, the first point to notice is that the usual effect of adding a filler is to increase conductivity, although it has been reported that glass fibre can have the opposite effect on nylon. It is interesting to speculate that since small traces of moisture can substantially increase conductivity, the use of a filler which is more hygroscopic than the polymer might at least initially have such an influence, and whether this was the case here. Of course, if this were the situation, it might well be that because of the attraction for moisture, this kind of filler would eventually lead to a contrary effect. In this connection, it is perhaps relevant to refer to a curious property of plasticised polyvinyl chloride in that addition of a filler such as talc or clay produces an increase in resistance, as does carbon also, but further addition leads to a sharp decrease. Particulate carbon is used in appreciable quantities to increase the electrical conductivity of natural rubber and synthetic polymers in general. With rubber, the effect becomes important at about 20 phr, the resistivity falling off rapidly with concentration (Fig. 4.3).[17]

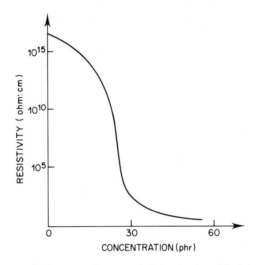

FIG. 4.3. Influence of filler on polymer resistivity (carbon black/natural rubber).

The shape of the curve depends on the surface and bulk nature of the carbon, but changes in conductivity from about 10^{-15} to 1 ohm^{-1} cm^{-1} are possible. Increasing the temperature can sometimes give rise to first a rise in resistivity and then a fall. This phenomenon has been reported for carbon in polystyrene and in polyvinyl chloride, and there has been speculation that the filler might exist in some extended aggregated form, like a necklace, depending upon fabrication conditions and that it is this continuity which assists high conductivity (see Fig 1.3). Heating to some critical temperature might cause sufficient expansion to break down the structure, thus isolating the conducting particles, and so overall, increasing resistance. Carbon-filled rubber also shows ageing effects which modify conductivity. Carbon and graphite fibres increase conductivity, aligned fibres producing the higher conductivity in the aligned direction as compared with the transverse direction, and continuous fibres having a greater conductivity than discontinuous fibres. Cross-ply systems which have no appreciable anisotropy and glass-fibre woven reinforced polyesters or random geometries do not vary very much in electrical conductivity with sample direction. The resistivity appears to decrease with thickness of laminate in paper–phenolic laminates and also, as expected, with humidity. Although, as indicated, carbon is the preferred conducting filler despite its pigmenting effect, because it causes less difficulty in processing and in deterioration of mechanical properties, considerable interest has also been shown in the use of metallic fillers. It has been shown, for example, that whereas nickel powder makes an appreciable contribution to conductivity below a concentration of 20 per cent for a polyethylene composite, some metallic fillers produce memory-switching effects at levels of about 55 per cent by volume[18] in which high conductivity under high-voltage conditions can develop and can be retained to low-voltage conditions. This can be removed by the passage of a high current when the composite returns to a high-resistance material. A so-called percolation theory, invoking particle–particle contact, can be used to understand this phenomenon. Again high temperatures can increase resistivity, presumably through a particle–particle separation effect. Coating of glass fibres with silver has been used to improve conductivity in unsaturated polyesters, parallelling at a higher volume fraction of course, the behaviour of silver itself. Oxidation of metal powder surfaces, thereby increasing surface resistance, reduces the influence of metal fillers. Flake and fibre metallic fillers further increase conductivity as do metal webs and fabrics. The conductivity of such structures can be expressed by the following equation:[19]

$$r/r_N = \frac{1-p}{1+11p^2} \tag{4.14}$$

where r and r_N are the resistivities of metal and metal network and p is a porosity parameter. These composites are also used for thermal and magnetic applications.

The resistance of a polymer or its composite to breakdown by high voltage expressed in terms of its dielectric strength, can be quite complicated involving not only direct material breakdown but failure due to tracking and arcing. Thus it will not only be a function of material properties but also impurities and voids. In general, it is found that fillers give rise to a fall in strength, e.g. a 30 per cent calcium carbonate concentration in a polyester resin reduces the dielectric strength from 440 volts/mil to 400 volts/mil although glass is reported to increase the value for nylon, polycarbonate, and styrene–acrylonitrile resin. If the filler is hygroscopic it will attract water into the polymer and by doing so, lower strength. Some inorganic fillers are believed to interfere with track formation and so can improve resistance to this form of breakdown, but concentration and the nature of the fabrication technique will further influence the behaviour. Fibrous fillers are reported to decrease discharge resistance.

Turning to *permittivity* and *dielectric loss*, it should be emphasised that both are affected by temperature and frequency, as indeed is dielectric strength. The most serious effect of adding a filler is through the production of interfacial polarisation at the interface, depending upon the ratio of permittivity of each component. Generally, mineral fillers and metals tend to increase overall permittivity whilst organic fillers cause a fall, but in both cases any tendency of the filler to attract moisture into the composite will modify the response. Attempts have been made to quantify the influence of concentration on permittivity and a number of equations have been proposed, with frequency also being taken into account. One equation, suggested by Nielsen[8] is, like earlier equations showing the general applicability of the principle, based upon the Halpin–Tsai modification of the Kerner equation. In the present context it takes the form:

$$\epsilon_c/\epsilon_m = \frac{1 + ABv_p}{1 - B\chi v_p} \qquad (4.15)$$

where

$$A = 1/A_e - 1$$

$$B = \frac{\epsilon_p/\epsilon_m - 1}{\epsilon_p/\epsilon_m + A}$$

and χ takes the form as in eqn (3.5). A_e is a depolarisation factor and ϵ_c, ϵ_m, and ϵ_p are permittivities of the composite, matrix, and filler.

Permittivity of glass-reinforced resins tends to increase with, for example, the value increasing from about 3 for a polyester resin to 4–5 for a chopped strand mat, sheet moulding compound, and fabric. The dielectric loss of a composite, expressed, say, as a function of frequency or temperature, differs from that of the unfilled resin, not only in the same way that mechanical loss does, but by the possible addition of an extra loss peak arising from interfacial, or Maxwell–Wagner polarisation. This is shown in Fig. 4.4. In the case of polymeric fillers, as occur in polymer blends, it is possible to obtain two maxima appropriate to both components.

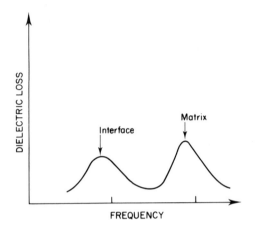

FIG. 4.4. Interfacial and matrix contributions to dielectric loss (schematic).

OPTICAL PROPERTIES

The use of fillers has an indirect effect, or direct in the case of pigments, on the optical properties of polymers, varying from complete opacity to some degree of translucency and colour formation. There are different ways in which for specification purposes, these properties may be categorised. For those polymers, such as polymethyl methacrylate, polystyrene, polycarbonate, and polyvinyl butyral, which are able to transmit up to about 92 per cent of visible light, the use of fillers must be carefully considered because of an immediate fall in transmissivity, although for aesthetic reasons in light covers, a certain amount of diffuseness may well be an advantage. Of course, reflection of light can take place for other reasons in polymers, such as because of crazes, cracks and, in crystalline polymers, spherulites. The basis of scattering is given by Fresnel's equation,[20] the form of which varies with angle of incidence and refraction. For the simple case of perpendicular incidence, it can be written:

$$R = \tfrac{1}{2} \frac{n-n'}{n+n'} \left(1 - \frac{n'^2}{n}\right) \tag{4.16}$$

where R is the reflectance and n and n' are the refractive indexes of the two phases in a composite. This equation, like its more general form, shows that the scattering, rather than the transmission of light depends upon the differences between the two refractive indexes. Thus, as will be seen from Table 4.3, a filler such as rutile will be the preferred filler for

TABLE 4.3
REFRACTIVE INDEXES OF FILLERS

Material	μ	Material	μ
Silica, E-glass	1·55	White lead	2·01
Calcium carbonate	1·57	Antimony oxide	2·09
Calcium sulphate	1·59	Zinc sulphide	2·37
Barium sulphate	1·64	Titanium dioxide (anatase)	2·52
Magnesia	1·74	Titanium dioxide (rutile)	2·76
Zinc oxide	1·99		

opacity, although anatase may alternatively be used, especially where the slight yellow cast of the former may be a disadvantage. It should be noted that the refractive indexes given in the table are the values against air, so that in a polymer, where the relevant differences are against the refractive index of the polymer, the absolute scattering may not be as high as in air, though the relative scattering between two different fillers can be very high. A filler with a refractive index near to that of a polyester resin, such as silica, can produce a translucent appearance in the composite. By careful choice of resin (Table 4.4), even this can be eliminated, at least at one temperature, although it would reappear on changing the temperature because of differences in the temperature coefficients of refractive index of the two components. Again, any void or crack formation which accompanies the fabrication of the polymer composite will add to the scattering of light. In a real situation, the inherent scattering may be further modified by the presence of a coating agent on the filler. If the filler has definite optical anisotropic properties, as is the case for calcite and talc, optical interference effects may produce colours in the composite.

TABLE 4.4
REFRACTIVE INDEXES OF POLYMERS

Material	μ
Orthophthalic polyester/styrene	1·57
Orthophthalic polyester/styrene (light stabilised)	1·56
Orthophthalic polyester/methyl methacrylate	1·55
HET acid polyester	1·55
Polycarbonate	1·59
Polystyrene	1·59
Polymethyl methacrylate	1·49

Particle size plays an important role in light scattering and indeed, this is the basis of a method for determining particle size, especially with the assistance of laser beams. When the size falls to the wavelength of light itself, then the composite may become transparent. For this reason, pigment manufacturers try to avoid the presence of fine particles, which may also produce colouring in addition to poor scattering. The reflectance expressed as a function of particle size is shown in Fig. 4.5. Of course, the situation in a real system may not be as bad as implied since agglomerates may form, although agglomerates and associated voids usually give rise to speckly scattering. It will be appreciated that in fairly concentrated systems, much of the scattering will be from multiple

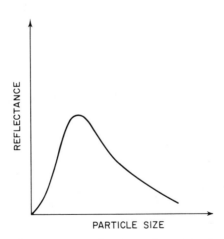

Fig. 4.5. Optical reflectance as a function of particle size (monochromatic radiation).

scattering involving repeated reflection from a number of particles. In addition to the other sources of scattering mentioned above for polymers, there is the added source in composites which may arise from imperfect bonding and from stresses within the system, and from any irregularities which the filler may introduce at the surface, again in addition to possible moulding defects. Before leaving this subject, mention should be made of the possibility of the filler affecting the crystallisation in crystalline polymers, since this would give rise to smaller spherulites and to a lower scattering, thus reversing the tendency normally associated with a filler. Talc can behave this way in polyethylene terephthalate.

MAGNETIC-BASED COMPOSITES

Although polymers have definite electric and other physical properties, they do not have an intrinsic magnetic character. However, polymers are used as vehicles for magnetic fillers, especially of the metallic ferrite type. In this way it is possible, by conventional moulding techniques, to produce both rigid and flexible magnetic components which are light in weight, electrically insulating, and easy to shape. Typical polymeric binders for this purpose are nylon, polyethylene, epoxy resin, polypropylene, and polyisoprene. These materials have found use in television, telecommunications in general, computer storage applications, and in automobile and other engineering. To illustrate the potential, with a 90 per cent level of ferrite it is possible to achieve a magnetisation of about 10^6 oersteds.

ACOUSTIC APPLICATIONS OF COMPOSITES

Concerning the use of polymers in the context of their acoustic properties, the main interest is in reducing the effect of sound. In addition to material properties geometric design of the system, involving decisions as to whether to use expanded or laminated plastics, is also important. Because of this and because frequency of sound, temperature, and air currents all play a part, the precise prediction of probable behaviour in a real situation is not always easy.

When a sound wave strikes a surface, part is reflected and part is absorbed, and even the behaviour of any which is transmitted will depend upon whether the sound energy was received by direct impact or whether it was airborne. Solid polymers have a high acoustic impedance compared with air and so are efficient reflectors of sound, although since the energy is mechanical in nature, if at a given temperature the polymer or polymeric component in a polymer blend exhibits a high mechanical loss for the frequency concerned, then absorption will occur. The approximate relationship which will allow extrapolation for frequency and resonance temperature is that the peak loss temperature increases by about 15°C for every decade increase in frequency. Thus a material like an elastomer with a low glass-transition temperature is a useful sound absorber at room temperature. Hard plastics on the same basis will give rise to greater reflection and thus a persistence of sound. Both polymer types are likely to be poor for reducing transmitted sound since this largely depends upon material density, but the use of sheet lead, as has been used with polyvinyl chloride, can assist sound deadening. Layered arrangements of graphite and mica flake have been found to increase the absorbing powers of natural and butyl rubber over a wide range of frequency. The effect of some fillers on acoustic resistivity is shown in Fig. 4.6.[21] A recent advantage claimed for polymers compared with conventional materials is in the use of reinforced polymers in water piping, where there is less 'flow noise' and presumably less tendency to 'water knock'.

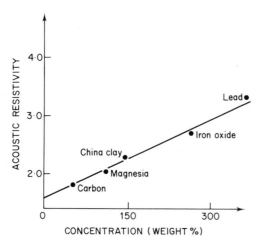

FIG. 4.6. Acoustic resistivities of vulcanised rubber composites.

FRICTION AND WEAR OF COMPOSITES

In many ways this topic should have been included under the heading of the mechanical properties of polymers, but as it has some rather special aspects, it is considered here. There are two main areas of interest. The first concerns the use of polymers and composites in low-friction applications as in bearings and gears, whilst the other is in high-friction applications such as in brakes and clutches, with tyres being an intermediate case, and a very important one. (The annual material wear of tyres is of the order of 10^6 tonnes.) In the above, low wear characteristics are generally desired except in the special case where high wear is tolerated because of high friction requirements in some single use applications.

Friction is concerned with the force which opposes movement between two surfaces and has elements of static, kinetic, and rolling friction. At a structural level there are associated elements of adhesion and material displacement, the latter arising from deformation and ploughing. The behaviour can be represented by Amonton's Rule, enunciated as long ago as 1699:

$$F = \mu W \qquad (4.17)$$

where F, μ, and W are the frictional force, coefficient of friction for a given material, and the load acting at right angles to the direction of the force necessary to overcome friction. Values of μ are given in Table 4.5.

TABLE 4.5
COEFFICIENTS OF FRICTION FOR VARIOUS POLYMERIC MATERIALS

Material	μ
Polytetrafluorethylene	0·25
Polytetrafluorethylene (pure)	0·05
Polyester	0·50
Polyester (30% carbon fibre)	0 28
Polyester (5% PTFE)	0·09
Polyvinyl chloride	0·45
Polyvinyl chloride (30% carbon fibre)	0·32
Polyacetal	0·21
Polyacetal (20% PTFE)	0·09
Polyethylene	0·68
Polyethylene (30% carbon fibre)	0·27
Polytetrafluorethylene (30% carbon fibre)	0·25

Although plastics obey Amonton's Rule fairly well, vulcanised rubbers do so less well, the coefficient of friction also typically increasing for plastics and decreasing for rubbers as the speed of sliding increases (Fig. 4.7). Movement of one surface on another generates heat, possibly by the alternate compression and relaxation of asperities, this being characterised by a PV term, where P is the normal pressure and V is the sliding

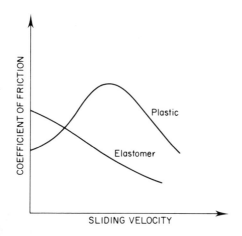

Fig. 4.7. Influence of sliding velocity on friction.

velocity. There is a limiting value of PV above which the temperature rises rapidly. In brakes this commonly means temperatures of about 200°C, but for vehicles on steep hills and for high braking conditions where heat is localised, temperatures of 2–3 times this are possible, leading to chemical breakdown of the matrix. Associated with friction is wear, with its elements of abrasion, erosion, fatigue, etc., so that the two properties of friction and wear may not necessarily be related. For instance, polytetrafluoroethylene has very low friction, but without the incorporation of a filler, it has low wear resistance. For this reason it might contain up to about 50 per cent of a filler such as asbestos, glass, bronze, or mica. There is a general relationship between hardness, ductility, and abrasion resistance, and use of a filler such as glass fibre stiffens a polymer to raise the PV limit. The hardness relationship to wear is shown in the use of fillers to reduce wear in floor covering, with irregular-shaped particles such as those of alumina, giving rise to faster wear than round particles. Other factors of importance in addition to concentration of filler, are particle size, orientation in the case of fibres, and adhesion between matrix and filler. Composites based upon phenolic and epoxy resins with fillers of cotton, asbestos, and glass have been used for many years in friction applications. Carbon black, especially channel black, is found to be useful for wear resistance in tyres, the performance rising to a maximum with increase of concentration. For low-friction applications, in addition to unfilled and filled polytetrafluoroethylene, other plastics such as nylon and acetal polymers also find use, possibly containing polytetrafluoroethylene itself as filler, or graphite, silicones, and molybdenum sulphide to improve friction and wear, particularly if reinforcing agents such as glass fibre or carbon fibre have been used to stiffen or strengthen the matrix, although carbon fibres can have a lubricating action in their own right. Advantages of polymer composites as bearing materials of this kind compared with other materials are that they do not necessarily need liquid lubrication and they are less prone to cold welding, as can happen with metals.

CHEMICAL PROPERTIES

Although the presence of fillers may have an influence on the preparation of polymers, talc for example causing a slight decrease in the rate of the free radical polymerisation of styrene, with zinc oxide giving a slight increase, and carbon behaving as an inhibitor, this section is mainly concerned with the chemical changes which the composite can undergo after preparation. However, perhaps mention should be made of the ability of fillers to reduce curing exotherms in, say, polyester cross-linking. Advantage of this is taken either to maintain a low temperature for reaction, or to permit the reaction to be carried out at higher temperatures, and so accelerate curing, than would otherwise be possible.

Chemical breakdown of a polymer can take place in a number of ways, both directly by the action of aggressive chemicals, and indirectly through stress-producing mechanisms. These include thermal, radiation, mechanical, and biological degradation. In practice, two or more of these may combine to create coupled breakdown environments. Examples of these are air, moisture, and sunlight in weathering or mechanical strain and solvent attack in environmental stress cracking. Sometimes these are generated at one and the same time; in other cases, attack first by one agent is followed by later attack by a second, and so on. There are also rate-determining features associated with breakdown, such as arise from rates of diffusion of chemical agent, generally increased by increase of temperature and geometric form of the moulded article, or reduced by use of a coating agent. In the context of this chapter it is not intended to deal with these in detail, but rather to highlight their importance in the design and use of composite systems.

Thermal degradation of polymers can take place by different routes, one or more being operative at one and the same time, and the relative preponderancy varying with the temperature. It may, for example, be through a depolymerisation or 'unzipping' mechanism, leading to virtual or complete breakdown of polymer chains, and often with the production of volatile low-molecular-weight material, which being organic in nature, frequently provides a foundation for inflammability. Other modes involve random breakdown of polymer chains leading to higher molecular weight products or to scission of side groups, again with the formation of volatiles and sometimes ring formation along the chain, the latter often being advantageous in that it can have some stabilising influence. Although not a great deal of information is available concerning the effect of fillers on thermal degradation, there is no doubt a body of knowledge in industrial laboratories which has not reached the usual scientific literature. However, it is known that certain metal fillers can accelerate degradation, presumably through a catalytic influence particularly where oxygen is present. For example, both copper and iron can shorten the induction period for significant breakdown, whilst glass may have no effect. In the case of epoxy resin degradation, the influence of filler increases with increase of nominal degradation temperature, with copper, iron, and aluminium having a positive effect, in that order.[22] For aluminium, the smaller the particle size, the greater was shown to be the effect. Calcium carbonate is reported to inhibit ethylene–vinyl acetate copolymer degradation, as does silica for a silicone elastomer. However, at high concentrations, above about 35 phr, the rate of degradation begins to increase again.[23] In general, if the matrix of a polymer composite is less mobile because of the use of a filler, then the diffusion of air and oxygen will be reduced, and this, like increase in vulcanisation in natural rubber, will in turn reduce thermo-oxidative degradation. If there is an increase in porosity, as is the case not only for expanded polymers, but through difficulties in compounding at high concentrations and with long fibres, and sometimes as a consequence of initial breakdown leading to loss of volatiles, then again degradation will increase for the reason implied above.

As has been mentioned, loss of volatiles can lead to inflammability in polymers and their composites, and for this reason certain chemicals are often added to polymers as flame retardants to reduce danger of fire. The addition of these may be at such a level that they constitute fillers, but quite separately certain fillers are also added to assist stability. The best-known additive is hydrated alumina, which in addition to its own non-inflammable character in diluting the polymer (perhaps at a level of about 40 per cent) has the advantage that at about 230°C it loses water by a cooling, i.e. endothermal, process. Being basic in nature also, it can interact with any hydrogen chloride which may be evolved with a polymer such as polyvinyl chloride. Calcium carbonate has the same property and can also provide a local blanket of carbon dioxide, and it also finds application for inflammability purposes. However, it should be pointed out that the temperature for massive loss of carbon dioxide is about 900°C, by which time most polymers with breakdown temperatures of about 300°C will have suffered extensive if not complete loss. The other important flame retardant filler is antimony oxide, which is generally used in conjunction with a halogen-generating compound with which it gives rise to synergistic behaviour. Two other particulate fillers used are zinc borate and hydrated sodium aluminocarbonate. Molybdenum compounds, such as molybdenum trioxide, have smoke-repressant qualities, and phosphorus compounds can produce an oxide glass on heating which plays a role in preventing further access of air.

The third area associated with thermal breakdown which is of special importance, is that of ablation, which has come into prominence in connection with space flight in which a returning space vehicle can develop external temperatures as high as 5000–15 000°C on re-entry to the Earth's atmosphere. Although these temperatures may obtain only for very short times, lower but, by normal standards, still very high temperatures of rocket engine exhausts at about 6000°C require longer applicational times. It is in this subject area, surprising as it may seem in view of the typical polymer degradation temperatures mentioned earlier, that composite polymer systems have played an important, if sacrificial, role, their advantages being shown in Table 4.6. The special advantage arises not only through good shock resistance, frequently a weakness in competitive ceramic materials, but through an ability of some polymers to produce a heat-resistant char, which although progressively eroding, retards penetration of a heat pulse towards the inside of the spacecraft. If a reinforcing filler such as an organic nylon or polyester fibre is used to strengthen the composite, its own breakdown leads to the production of

TABLE 4.6
ADVANTAGES OF ABLATIVE POLYMER COMPOSITES

Low thermal diffusivity
Evolution of cooling gases
Tendency to char rather than melt
Good shock resistance
Low density

volatiles which can cool the charring thermosetting polymer matrix, and thus further add to the heat resistance. For less aggressive conditions which may warrant the use of mechanically superior fillers, glass, carbon or boron fibre combined typically with an epoxy resin are often used.

Although biological breakdown is not a serious problem with synthetic polymers as it may be with natural polymers, it can be of importance where natural polymeric material is used as the filler, attack coming from bacteria, fungi, insects, and even rodents, leading to mechanical failure of the composite. Indeed, this tendency has been exploited by Griffin[24] to produce disposable plastic bags in which a biodegradable filler, such as starch, in an otherwise inert matrix, like polyethylene, will eventually break down leaving a friable polymer matrix which is readily broken down mechanically, and is easily dispersed without detriment to the environment. Implicit in the breakdown process is the ability of the starch products to be leached out, which brings us to another aspect of composites in the context of degradation. This is the permeation of liquids in particular, through polymer matrixes.

Permeability is a property which has important consequences, say, in the attack of composites by chemicals and especially water as we shall see. On the other hand it has the advantage of controlling the supply, say, of drugs or other medicaments, which may be incorporated into a polymer for treatment of disease. In general, the permeability of polymers can either decrease or increase when fillers are used, depending on such factors as concentration of filler, particle size, shape, distribution, and binding between filler and matrix. It will also depend upon the chemical natures of filler and matrix. Whereas, other things being equal, increase of concentration should cause a decrease in permeability, any agglomeration, crazing or crack-formation associated with the use of the filler, and even brought about by the presence of the diffusant itself, can lead to an increase in the rate of diffusion. If the diffusant can in turn actually attack the filler then transport will be further aided. If the attack is alternatively at the interface between filler and polymer, then passage around the filler can increase diffusion, assisted as it might be, by capillary action. Use of some suitable coating agent, such as a silane on glass, can reduce or prevent this. Many polymers can undergo breakdown by attack by ultraviolet radiation and other high-energy radiation. Lead-filled polymer has been used to combat the latter, but the former is of more general concern because of the use of polymers and composites in outdoor applications. For this certain additives, in conjunction perhaps with UV chemical stabilisers, are often employed. These might, typically, either absorb or reflect radiation as their main role. The additives which are used in this way include carbon, which is particularly active over a wide range of radiation frequency, zinc oxide and titanium dioxide, rutile being more effective in this respect than anatase and finding application in polyvinyl chloride flooring tiles, etc. Some inorganic fillers can re-transmit UV radiation as visible light, producing fluorescence.

Mechanical degradation is another mechanism by which polymers may be both chemically degraded or modified, and takes place when the rate of stress build-up exceeds the ability of a bond to respond by translation, and therefore scission occurs. From this formation of a free radical, a number of possible paths are available for the next stage, depending upon the stress conditions, temperature, and local environment. There are basically two ways in which action arises. For rigid or brittle polymers in which the strain is small, brittle fracture, with or without chain fracture, may take place, but for softer but highly viscous polymers, about which most is known regarding mechanical degradation, an inability of a bond to deform fast enough to shearing forces is the origin. Since rigid particulate fillers increase the tendency to brittle fracture, their presence can increase the sensitivity of the polymer to this type of breakdown. Where there is strong intermolecular interaction between the filler and the polymer chains as in carbon–reinforced natural rubber, localisation of some chains by the filler more than others is believed to be a cause of strain softening, in which the composite becomes less stiff on working, a phenomenon known as the Mullin's Effect, and attributed, at least in part, to the progressive breakdown of polymer chains (Fig. 4.8).[25]

Although sensitive to the attack of specific chemicals, many polymers have advantages over other conventional materials of construction, which has led to their use in chemically aggressive environments. At the present time, the loss in steel through corrosion amounts to many

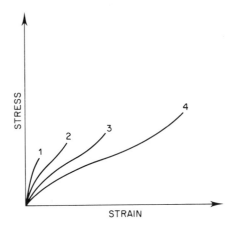

FIG. 4.8. Strain softening in a carbon black/SBR composite. (Strain sequence 1 to 4.)

millions of tonnes each year, and in the USA it has been estimated that 60 per cent of steel goes into replacement of existing structures. No wonder, therefore, that polymers are finding wide application, especially in marine applications, with excellent resistance to acids and many other chemicals, although the largest tonnage reinforced polymers, the poly-esters, do suffer from a susceptibility to alkali attack. In the replacement of steel structures by plastics, there is another advantage of reinforced plastics in that it is relatively easy to make seamless mouldings and therefore remove one of the usual drawbacks of some structures, like ducting and tanks. In addition because of their long life and durability, the cost of maintenance is low and repair is relatively easy when required. When attack does occur, any failure of the composite can involve breakdown in the resin, or in the filler and at the filler–resin interface. To minimise the last two possibilities it is important to ensure as far as possible, that the outside of the moulding is composed of resin, even if the final article needs to be given a protective covering, or gel coat.

Being organic chemicals themselves, the resins may be attacked by organic solvents which may swell, dissolve or cause stress-corrosion (see below) of the system. For polymeric fillers the same may also occur. But of all the liquids with which the composite may come into contact, the one which has given rise to the most concern in view of its universal occurrence and usually unfavourable influence, is water. It is appropriate, therefore, at this stage to say something about the influence of water on polymer composite properties, and in particular on the mechanical properties, since the effect on electrical properties will be self-evident as will be that on the optical properties where any debonding is likely to produce light scattering and opacity, as indicated by eqn. (4.16). The vulnerability of polyester–glass fibre composites was appreciated in the very early days of their manufacture and was found to be aggravated by temperature rise. It was this vulnerability, rather than any hoped-for improvement in mechanical dry strength, etc., which probably stimulated the research into coupling agents for composite polymer systems. In fact, composites as a whole are generally sensitive to the presence of moisture, and the further influences of temperature and time of contact as well as such factors as the nature and magnitude of any stress, the lay-up or geometrical organisation of filler, and of course the actual nature of the filler and any pretreatment which it may have had, and the general chemical nature of the resin, will all be important. It is of interest from the applicational point of view to be able to assess the long-term behaviour of composites and since temperature and residence time are relevant parameters, as indicated above, much use is made of the 'boiling water' test in this connection, particularly in comparing the effectiveness of different coupling agents. In this way it can be shown that an unsized or heat-cleaned glass fibre in an epoxy or polyester resin will produce a composite in which tensile strength, shear strength, and flexural strength (this in particular being very sensitive to boiling water treatment) all suffer serious deterioration. On the other hand fibre treated with coupling agent may only lose perhaps 4 per cent in tensile strength. The effect of temperature alone, and temperature and moisture is shown in Fig. 4.9 for a boron–epoxy composite.[26] Again a combination of moisture and applied stress can also lead to early breakdown. For cross-ply glass-reinforced polyesters, a rapid take-up of water is found to occur at about the knee position in the stress–strain curve (Fig. 3.13) and it is believed that water enters along cracks, the sudden uptake not occurring with unidirectional aligned fibre composites, which do not exhibit a 'knee'. For laminates in which fibre ends are exposed, earlier breakdown takes place than for systems with sealed ends. It is probable that a good laminate of this kind, finished with a gel coat, would have a lifetime in a weather environment like that of the UK, of over 30 years. Different types of filler will have different resistances to attack, and although the nature of the coupling agent will be important, it has been reported that a carbon fibre behaves better than a boron fibre in a polyimide matrix.

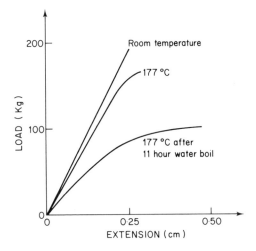

FIG. 4.9. Influence of temperature and boiling water on the load–extension properties of a boron fibre/epoxy resin composite. (Reproduced with permission of American Chemical Society.[26])

From the resin point of view, polyesters prepared from isophthalic acid are superior to those prepared from orthophthalic, and a bis-phenol-A-based system better than either. Concerning the mode of breakdown under the influence of moisture, there is no doubt that ingress can occur by wicking along fibres, but it appears that there is no single mechanism for breakdown in all situations. Further, whilst some workers have found that the failure is reversible, in that the strength properties are recovered on drying, other workers report irreversible damage, no doubt the different conclusions arising, in part, from different mechanisms of breakdown in the first place. Finally, with reference to other liquids, oil has been found to have little effect on the torsional fatigue of carbon-reinforced composites, whilst benzene causes earlier failure in glass–epoxy systems, the failure not being due to swelling as might be thought, but to a fall in the surface energy for flaw growth, or fibre–matrix debonding.

The role of organic liquids is well known in another connection regarding failure of polymers and their composites, through what is called environmental stress-corrosion or solvent crazing. This takes place through contact with certain organic liquids, which produce premature failure in stressed materials, the stress being deliberately applied or inherent from some limitation of the fabrication technique. The special influence in the case of composites arises possibly from stress fields near to the filler and due to the presence of filler, or from differential cooling effects of polymer and matrix. It has already been shown that more than one degradative influence may be operating at one and the same time. Perhaps a good illustration of this is in 'weathering' of composites. Because of its importance in outdoor applications of polymers, special xenon-arc artificial weathering equipment is used to simulate accelerated weathering. The overall influence will embrace the effect of heat, sunlight, ozone, moisture, and other common aggressive chemicals such as carbon dioxide and sulphur dioxide, and so it is very difficult to reproduce a typical exposure situation, which may further be complicated by dust, dirt, fog, etc., as well as the different shapes and constructions which a polymer composite will find itself in. However, some general conclusions may be reached as to how a system is likely to respond to a weathering situation. For instance, mechanical properties tend to deteriorate with a greater change occurring in flexural behaviour than tensile, although in the short term there may be an improvement in properties, presumably arising from some post-curing reaction brought about by light energy, or heat if the original cure was at room temperature. Optical properties typically change by yellowing, loss of surface gloss and by an increase in opacity sometimes preceded by the appearance of the fibre weave pattern near the surface. Nevertheless, compared with many other materials, reinforced plastics have demonstrated excellent weathering properties and are now first choice for many outdoor applications.

REFERENCES

1. See HOLLIDAY, L. and ROBINSON, J. D., in Richardson, M.O.W. (Ed), *Polymer Engineering Composites*, Chapter 6, Applied Science Publishers, London (1977).
2. GRESZCZUK, L. B., *Soc. Plast. Ind. 20th Ann. Meeting of R.I.P. Div. Proc.* Section 5C, p. 10. (1965).
3. SCHAPERY, R. A., *J. Comp. Mat.*, 23, 80 (1968).
4. LUBIN, G. (Ed), *Handbook of Fiberglass and Advanced Plastics Composites*, Van Nostrand Reinhold, New York (1969).
5. HANDS, D., R.A.P.R.A., Shawbury, private communication.
6. ISHIKAWA, T., KOYAMA, K. and KOBAYASHI, S., *J. Comp. Mat.*, 12, 153 (1978).
7. MAXWELL, J. C., *A Treatise on Electricity and Magnetism*, Clarendon Press, Oxford (1904).

8. NIELSEN, L. E., *Ind. Eng. Chem. Fundam.* **13**, No. 1 (1974).
9. SUNDSTROM, D. W. and CHEN, S. Y., *J. Comp. Mat.*, **4**, 113 (1970).
10. CHENG, S. C. and VACHOU, R. I., *Int. J. Heat Mass Transfer*, **12**, 249 (1969).
11. ZIEBLAND, H., see ref. 1, Chapter 7.
12. KNIBBS, R. H., BAKER, J. D. and RHODES, G., *26th Am. Tech. Conf.*, R.P.C. Div. Soc. Plastics Ind. (1971).
13. BIGG, D. M., *Composites*, **10**, 95 (1979).
14. NOGA, E. A. and WOODHAMS, R. T., *Soc. Plastics Eng. J.*, **26**, No. 9, 23 (1970).
15. TITOV, W. V. and LANHAM, B. J., *Reinforced Thermoplastics*, Applied Science Publishers, London (1975).
16. SCARISBRICK, R. M., *J. Phys.* **D6**, 2098 (1973).
17. NORMAN, R. H., *Conductive Rubbers and Plastics*, Elsevier, Amsterdam (1970).
18. ARAIJO, F. F. T. DE, GARRETT, K. W. and ROSENBERG, H. M., *Proc. 1975 Conf. on Comp. Mat.*, Met. Soc./A.I.M.E., New York (1976).
19. KATZ, H. S. and MILEWSKI, J. V., *Handbook of Fillers and Reinforcements for Plastics*, p. 610, Van Nostrand Reinhold, New York (1978).
20. See MASCIA, L., *The Role of Additives in Plastics*, Chapter 5, Edward Arnold, London (1974).
21. HATFIELD, P., *Brit. J. Appl. Phys.*, **1**, 252 (1950).
22. SHELDON, R. P. and LUNN, D. J., unpublished results.
23. MIRAKHMEDOV, M. M. and AKUTIN, M. S., *Plast. Massy*, No. 3, 7 (1977).
24. GRIFFIN, G. J. L., in Deanin, R. D. and Schott, N. R. (Eds), *Fillers and Reinforcements for Plastics*, A.C.S. Adv. in Chem. Ser. 134 (1974).
25. MARK, H., GAYLORD, N. G. and BIKALES, N. M. (Eds), *Encyclopedia of Polymer Science and Technol.*, vol 12, p. 48, Interscience, New York (1970).
26. BROWNING, E. and WHITNEY, J. M., in Deanin, R. D. and Schott, N. R. (Eds), *Fillers and Reinforcements for Plastics*, A.C.S. Adv. in Chem. Ser. 134 (1974).

Chapter 5

RHEOLOGY, COMPOUNDING, AND PROCESSING OF COMPOSITES

FLOW BEHAVIOUR OF POLYMERIC SYSTEMS

The essence of processing polymeric systems is to be able to establish conditions such that the polymers are easily deformed whilst they are being shaped but once this has been achieved, the shape is then retained for whatever is the chosen application of the material. The shaping embraces the elements of flow and time, and in a very general sense, the aim is to increase the former at the expense of the latter, provided that there is no serious effect on ultimate properties. In this way, fabrication takes place as quickly as possible, i.e. cycle time is short, thus minimising energy and cost and possible deterioration of the matrix and filler. The shaping sequence can be carried out by melting and softening of the polymer, moulding and then cooling, or by reaction of relatively low-molecular-weight and mobile reactants in a mould, or by evaporation of solvent or suspending medium in the case of an emulsion or dispersion. For filled polymers, in addition to the forming process, there is a previous stage of compounding, i.e. mixing and dispersion of filler. In both stages of the overall process, perhaps with the exception of dry mixing, flow of polymer or polymer solution, gel, etc., is fundamental to processing, and so a discussion of the flow behaviour of polymers and their composites is relevant.

Except at extremes of high pressure or low temperatures, low-molecular-weight fluids exhibit Newtonian flow behaviour, i.e. the rate of deformation (flow) is proportional to the applied stress:

$$\tau = \eta \cdot \frac{d\gamma}{dt} \qquad (5.1)$$

where τ is the shear stress, $d\gamma/dt$ is the strain rate of flow, and η is the coefficient of viscosity. As the molecular weight increases as, say, with chain length of a polymer, this simple relationship begins to break down not only in the melt but in solution and gel also. In other words, in homogeneous liquids, the flow behaviour becomes non-Newtonian. In principle, this deviation can take a number of forms as shown in Fig. 5.1,

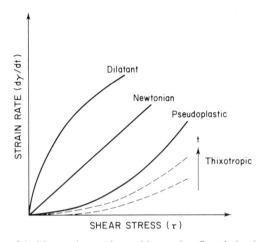

FIG. 5.1. Newtonian and non-Newtonian flow behaviour.

that of the pseudoplastic flow approximating to general polymer tendency. For some filled systems in which a loose structure of filler develops, often produced by clay systems in which interaction exists between particles, a time-dependent breakdown of structure takes place, known as thixotropy. On removal of stress the quasi-structure reforms with the apparent viscosity increasing, a phenomenon specially designed to operate in some surface coatings in order to reduce the tendency of a paint to run. The non-Newtonian behaviour is often reflected also in an increase in elastic behaviour of fluids, contributing to overall viscoelastic behaviour (see Chapter 3).

Different polymers behave in different ways within the framework of pseudoelastic response and can, to a first approximation, be represented by a constant, n, in a Power Law relationship:[1]

$$\tau = K \cdot (d\gamma/dt)^n \qquad (5.2)$$

where K and n are constants for a particular system at a given temperature. Various factors will affect the flow behaviour, notably average molecular weight, molecular weight distribution, chain branching, and chemical nature. It is not intended to go into these in detail here, but in general it can be said that the more polar the polymer, i.e. the greater the interaction between molecules, the higher will be the viscosity. As far as molecular weight is concerned, above a critical molecular weight which is associated with onset of chain entanglement, the equation linking viscosity to molecular weight takes on a form which applies to many polymer melts and concentrated solutions:[2]

$$\log \eta = k M^{3.4} \tag{5.3}$$

where k is a constant for a given polymeric system at a given temperature and M is average molecular weight. The critical molecular weight depends upon the degree of branching, as does k also. Under very high shear rates, the shape of the basic molecular flow unit, assumed to be a limited number of chain segments, can change, becoming more oriented in the direction of flow, and in so doing can produce a change in the viscosity behaviour. If the rate of stressing becomes too high, the molecule may develop local stress leading to chain fracture and, again, to a change in the basic flow character because of mechanical degradation. Temperature will usually increase the deformation rate for a given stress, although for the special case of a highly viscous elastomeric system, the actual deformation may, in fact, be more restrained. The influence of temperature not too far above the glass-transition temperature is given by the Williams–Landel–Ferry (WLF) equation[3] (eqn (5.4)) and for higher temperatures, say about 100°C above the glass-transition temperature, or for low-molecular-weight polymers, by the Andrade equation,[4] eqn (5.5).

$$\log \eta / \eta_g = \frac{a(T - T_g)}{b + T - T_g} \tag{5.4}$$

$$\eta = A e^{E/RT} \tag{5.5}$$

where η and η_g are the viscosities at temperatures T and T_g (glass-transition), a and b are universal constants, A is a constant and E is the activation energy for viscous flow. To give some idea of the influence of temperature in practical terms, a decrease in temperature of about 30–50°C for flow of a polymer through a die would require typically an increase in the pressure head of a factor approximately 10^3 atmospheres to maintain the same rate of flow. Highly viscous polymer liquids can develop quite strong elastic responses, illustrated by the well-known

Weissenberg effect,[5] involving normal-stress effects, such as a tendency for the polymer to flow up the stirrer shaft rather than away from the blade in propellor stirring, or for some so-called elastic liquids to apparently flow against gravity when the stream of normally flowing liquid is interrupted. The same properties are usually demonstrated in the form of die-swell in extruded polymer in which the extrudate diameter is larger than the die diameter, and as melt fracture associated with, e.g. 'sharkskin effects' in the solidified material, especially for the case of high cooling rates. Similarly where there are significant differences in cooling rates between the surface and interior, a not uncommon situation for materials of such low thermal conductivity, different morphological and molecular arrangements, together with voiding, can arise across the axis of the extrudate. It will be appreciated that factors like melt temperature, specific heat, and any heat of crystallisation will all have an influence and in turn will affect cycle time for optimum processing. Die swell can be reduced by a lower strain rate, increasing temperature and decrease of molecular weight, and, as we shall see, by the use of fillers.

In general, the complications of polymer flow compared with that for simple systems, will tend to be further aggravated by the presence of fillers. The immediate effect of a filler is, without specific reference to chemical nature and shape, to increase viscosity, interfere with the polymer flow pattern in a particular process, produce thixotropy, and to give rise to machine wear. They can also, as we saw in the last chapter, affect the thermal properties of the system by, for example, reducing exotherms and by conducting heat away from the matrix more rapidly than before, both influences giving rise to shorter cycle times. However, the most important effect of fillers is on viscosity and since this is an important design parameter for processing, let us look at the various ways in which a filler can modify viscosity. The relevant properties of the filler are concentration, size, aspect ratio and shape, fibre length for fibres, stiffness, strength, and the specific interaction, related to wetting, which can exist between filler and matrix. Within these are the special factors of easily deformable fillers and those whose viscosity is not dissimilar to that of the matrix, as in polymer blends and emulsions, and which can lead both to change of shape and breakdown of filler. A cautionary word of warning should be given here: not all the different methods of measuring viscosity may give the same results because of these factors, so care must be exercised in using data obtained under one set of flow conditions for application under another.

The starting point for considering the first of the above parameters, concentration effects on the viscosity of dispersed particulate systems, is Einstein's equation[6] (eqn (3.2)):

$$\eta = \eta_1 (1 + k_E v_2)$$

For rigid spheres and normal flow k_E takes the value of 2·5; for fibres the value increases with aspect ratio adopting the value of $2 \times$ aspect ratio for uniaxially oriented filler, in both cases for dilute dispersions. At higher concentrations where particles begin to interact with each other, the Einstein equation is no longer applicable and a number of modifications or empirical equations appropriate to some specific system have been introduced to describe this deviation. The simplest is that of Simha and Guth,[7] which is the first stage of a polynomial modification:

$$\eta = \eta_1 (1 + k_E v_2 + 14 \cdot 1 v_2^2) \qquad (5.6)$$

It has been reported that for systems of carbon black and calcium carbonate with polyisobutylene and butadiene elastomers, Einstein's equation is applicable up to 10 per cent concentration after which the Guth–Simha equation gives better agreement. Of other equations mention might be made of those of Cross,[8] Maron-Pierce,[9] and Mooney[10] (eqn (3.3)). It is frequently found that the flow curves of filled polymers are superimposable at a fixed stress, provided that the aspect ratio is small. A certain amount of theoretical and experimental effort has gone into studying the effect of particle shape and a number of equations, typically of a polynomial form with respect to concentration, have been cited, but for our purposes those relating to systems of fibrous fillers are of most interest. In principle, for dilute systems, the appropriate value of the Einstein coefficient should suffice. However, since this depends not only on aspect ratio but also on orientation with respect to the stress direction, beyond giving higher viscosity values than for fillers of low aspect ratio, it is not easy to predict the absolute effect. It might be mentioned that for the same aspect ratio, flakes give even higher viscosity changes.[11] With fibrous-filled systems, increase of concentration leads to a loose network formation moderated by the presence of any dispersing aids but showing elastic behaviour, thixotropy and in particular, a yield stress. In short, the composite develops pronounced non-Newtonian properties, such as, for example, the pseudoplastic nature of polypropylene containing glass fibre. This non-Newtonian behaviour is not confined to fibre-filled systems, however, phenolic and polyester resin systems which show Newtonian characteristics soon becoming non-

Newtonian with the addition of mineral fillers. Carbon-filled polyiso-butylene also exhibits a yield stress the system behaving as a Bingham solid, presumably because of the active surface of the carbon which will lead to agglomeration in the absence of stress.[12] Any tendency of all systems in this direction will depend not only on concentration but on particle size, since a decrease in this will lead to a greater specific area for the filler. The interaction giving rise to agglomeration can be assessed to some extent from oil absorption data, but typically a polar filler dispersed in a non-polar matrix will be prone to agglomeration, and for this reason may well require the addition of dispersing agent to the system, or suitable coating of the filler. It will also need longer mixing times for breakdown during compounding and will be less in evidence for high-shear as distinct from low-shear, conditions, whether brought about by adjustment of matrix viscosity or processing technique. Temperature increase will also generally assist breakdown. There has been some interest in the flow behaviour of mixed filler systems, either of particulate fillers of different sizes, or of particulate–fibre mixtures. To a first approximation, the viscosity of the mixture can be expressed as:[13]

$$\eta = \eta_1 \; \Pi_i \; (1 + k_E v_{2_i}) \qquad (5.7)$$

where v_{2_i} is the concentration of filler species, i. A related equation with an extra filler–filler interaction term has been suggested for fibre–particulate systems. The theories of flow in mixed systems predict a minimum in the viscosity-relative concentration curves;[14] this has been shown to occur for mixed quartz and marble-filled polyester (Fig. 5.2).

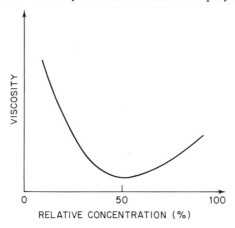

FIG. 5.2. Viscosity of liquid containing mixed particulate fillers.

As already indicated in connection with agglomeration, increase of shear rate will affect the flow behaviour of composites, with usually the apparent viscosity decreasing with increasing shear rate. This, however, may not altogether be as a result of a change of flow pattern, since both polymer matrix and filler may suffer breakdown, the former by mechanical degradation and the latter by attrition. For example, the length of fibre can be reduced by a factor of three in high-shear processing of glass-reinforced polyethylene. For low rates of shear, fibre axes will be oriented at about 45° with respect to the overall flow direction, but with increasing shear rate, the axes will turn more and more into the direction of flow. On removal of stress the molecules will tend to recover more quickly than the fibres, and thus a permanent orientation effect, well recognised in fibre-reinforced thermoplastics, is retained. Another organisational effect may arise from the nature of the flow profile across the die or tube (Fig. 5.3) in that rigid fillers tend to migrate to a position halfway

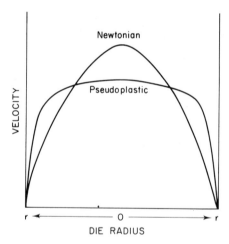

FIG. 5.3. Flow profiles across a die for Newtonian and pseudoplastic systems.

between the wall and the central axis for Newtonian liquids undergoing laminar flow, and towards the centre for non-Newtonian liquids. This influence could affect mechanical properties, although it would only become really significant in narrow geometries. Flow for concentrated systems in narrow dies can lead to another complication comparable to a process of 'log jamming' when the fibre diameter is of the order of the die diameter, again emphasising the equipment influence on the value of measured viscosity. Die-swell with filled systems, whether the fillers are rigid, such as mineral fibres, or even flexible as in polymer blends, tends to be reduced, although any breakdown of filler, including dispersion of agglomerates, reduces the effect. High concentrations of fillers can also lead to surface rupture, thereby producing irregularity of profile.

So far discussion has been almost entirely on polymers which contain non-deformable fillers. A class of polymer composites which at one extreme coincides with this description, but at another may, as indicated above, contain a deformable filler, is a polymer blend, the transition between the two limits being brought about typically by change of temperature. Such systems are further complicated by changes in concentration which can bring about phase inversion depending upon the relative fluidity of the two components at the moulding temperature. Alternatively, again depending upon temperature and concentration, both phases may be continuous, or one component may be dispersed in another with a third, continuous, phase being itself of a different composition. It will be realised from this that in terms of their rheological properties, polymer blends are quite complicated systems. For dilute compositions, an Einstein relationship (eqn (3.2)) might very well apply, but as the shear rate is increased the filler particle will begin to elongate and either break up into smaller particles or remain elongated as a quasi-fibre. Whatever the situation, the viscosity changes are far from regular as shown by Fig. 5.4.[15] Where morphological transitions such as the

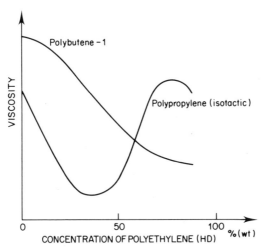

FIG. 5.4. Viscosity of polymer blends.

above do take place, although they may be retained upon cooling of the blend, any subsequent reheating or relaxation brought about by solvent action could result in a further change in morphology, and therefore in mechanical and physical behaviour. For general prediction of flow characteristics, a logarithmic Rule of Mixtures has been proposed:[16]

$$\log \eta_c = v_1 \log \eta_1 + v_2 \log \eta_2 \qquad (5.8)$$

where 1 and 2 refer to the separate components. In the light of the above complications, however, this equation can provide no more than broad guidance.

Finally, with reference to temperature effects on the viscosity of filled polymers, it should be mentioned that Andrade's equation (eqn (5.5)) has been reported to apply fairly well to dispersions, surprisingly as diverse as a 40 per cent calcium carbonate in polypropylene composite and a 25/75 polyethylene/polystyrene polyblend.

MIXING AND DISPERSION

One of the most important stages in the preparation of polymer composites is the process of distributing the filler in an even manner throughout the polymer matrix. By doing this, any tendency to form a gradient in composition and properties is reduced, at least to a level acceptable for the desired application of the composite. In this section we shall consider some of the principles involved in doing this, and then turn to the actual processes for achieving the distribution. Although the subject is not altogether fully understood, in the sense that a great deal of empiricism is involved, it is of great technological importance in terms of equipment and fuel costs. The actual choice of distribution technique will depend therefore on cost and convenience but above all on the state of the polymer and filler. Typically it will be one of direct mixing for solid suspensions, high-viscosity shear mixing for many polymer–filler systems, paste working, and for some cases particularly appropriate to some polymer blends, by dissolution techniques. The overall process of distribution is actually composed of two elements, true mixing and dispersion; these will now be discussed.

Mixing concerns the alternation in space of two components (if there are more than two, the same treatment can still be imagined, with all but one component being regarded as an entity, and the excluded material being the second component). The process of mixing is carried out until an acceptable statistical distribution is achieved. The principles involved in liquid manipulation will depend on whether the liquid is behaving as a Newtonian liquid when turbulence, eddy, diffusion, and laminar mixing may all be taking place, or as a non-Newtonian fluid when laminar mixing will predominate. In practical terms, the latter can mean that mixing periods perhaps fifty times longer than for Newtonian liquids are necessary. Complications can arise when the process changes the nature of the system from one type to the other. To describe the extent of mixing, two divisions of scale and intensity are invoked. The scale of mixing is a measure of the average distance between centres of maximum difference in properties, and intensity of segregation is a measure of the average deviation of the concentration at a point from the mean concentration. It can be shown that the larger the initial scale of segregation and the smaller the ultimate scale of striation thickness the more work must be done, which although perhaps self-evident, indicates the possible value of some pre-preparation of components before the main mixing process is begun. The same is true for small amounts of additive in that the smaller the quantity of filler, the more physical mixing must be carried out to obtain a suitable distribution. It is for this reason that the technique of 'masterbatching' is commonly used in polymer compounding. In this a concentrated mix of polymer and filler is used in mixing, rather than adding the small quantity of filler to the polymer in the first place. The acceptable level of scale and intensity depends upon the particular system—a mixture does not have to be completely random to be suitable for a given purpose. Indeed if mixing is a prerequisite to fabrication rather than a simultaneous process of mixing and moulding, there is always the possibility of some demixing occurring in the moulding, perhaps because of differences in temperature and mechanical stresses not present in the original mixing. Tendencies towards aggregation because of changes in the relative compatibility of filler and polymer may also be present. Thus excessive time and expense associated with mixing is unnecessary. To perhaps highlight the difference between mixing and dispersion it may be pointed out that the latter is specifically concerned with the breakdown of what may be states of agglomeration or association, which becomes increasingly difficult as particle size decreases, particularly at the 0·2 micron level. The difficulty varies from class to class with, say, carbon black and iron oxide being more troublesome than titanium dioxide, and polar polymers aiding dispersion more than do non-polar polymers. It is in situations like this that dispersing agents become important as illustrated by the fact that whereas the normal incorporation time for substantial amounts of zinc oxide in polyethylene is about ten minutes, with the use of a hydrophobic surface treatment, the time is reduced to about three minutes.[17] Dispersion is made up of the three stages of initial wetting, breakdown of agglomerates, and then intimate wetting of particles to displace air pockets. It is found that the agglomerates are best broken down when the shear forces are greater than some critical value, which varies from system to system, the shearing action bringing a constant supply of fresh polymer into contact with the broken-down particles. The dispersion becomes more and more difficult as the process goes on, following a roughly logarithmic relationship with time and, for a particular system, approaching a limit to the degree of dispersion. Large agglomerates are more easily dispersed than small so that again, use of a concentrated initial mixture, say about 70 per cent by volume, is desirable.

The question now arises as to how one assesses the efficiency of a particular mixing and dispersion process, obviously an important factor in the choice of a preferred distribution method. The most common technique is the visual one, possibly extended by using the microscope or electron microscope, etc., but as far as the eye is concerned, being capable of distinguishing between sizes of the order of a thousandth of a centimetre. Specks, streaks, and spots are the usual criteria. Alternatively, use may be made of some other physical property, particularly a mechanical property such as tensile strength or elongation to break, especially if the quality of the material for user application is going to be assessed by the same property (Fig. 5.5).[18]

COMPOUNDING AND FABRICATION OF COMPOSITES

Compounding

Reference has already been made to compounding as an essential step in the fabrication of good quality products, preparing the polymer system for its final moulding procedure. The product of the compounding stage, unless an integral part of the moulding process itself, can be in many forms appropriate to the moulding requirements, examples being powder, slurry or paste, ropes, premixes, chips and other particulate forms, and preforms. This step is often carried out by the material supplier, but may be done by specialist organisations or, for the larger concerns, by the fabricator. Thus, although the precompounded material is more expensive than direct mixing, there is the cost of the compounding

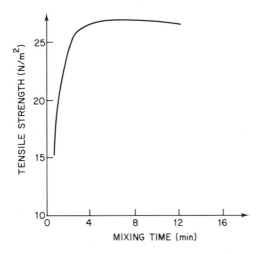

FIG. 5.5. Influence of time of mixing on tensile strength of SBR/carbon black composite.

equipment, storage space, etc., to be taken into account. As mentioned earlier, the theory of mixing, complicated by the fact that in a real system there may be many ingredients other than filler to be considered, is generally not adequate for predicting the effort needed for optimum compounding conditions. Because of this, trial and error assessment, perhaps aided by use of a torque rheometer in some cases, is generally employed.

The compounding procedure is basically in three steps which may all be done in the same apparatus or in separate machines. There is the premixing stage for the breakdown, then dispersion of filler, possibly assisted by external or internal (work) heating, and then a final shaping stage in which the composite is converted, for example, into rod, for pelletising, or into rope or tape. For solid compounding the three basic techniques are: internal mixing, milling, and extrusion, but in addition there are a number of miscellaneous approaches such as pultrusion, filament winding, spray mixing, and so-called motionless mixing. Before material is introduced to one of the above processes some dry mixing of the ingredients may be carried out. Alternatively, for emulsions, components can first be mixed directly, due care being taken to ensure that the nature of the stabilising agent for each is the same, since otherwise coagulation would occur, or if one component, say the filler, has not been given a treatment already, then to make sure that the correct type of agent is used. If dry mixing is first carried out, it must be done in full recognition of the next processing stage. For example, although dry blending would be satisfactory for material which is then transferred to an extruder or screw injection moulding machine, since both involve further mixing, it would probably not be suitable for compression moulding or ram injection moulding in which no further mixing is available. Also as for any mixing technique, care must be taken to prevent excessive build-up of heat with phenolic and other thermosetting systems, so that no undue premature reaction takes place.

Compounding through the use of an internal mixer is achieved primarily by the shearing action of the blades, the equipment usually being based upon a Banbury mixer which has two connecting chambers, each with a Z-blade turning on rotors so that the blades intersect with each other. There is provision for heating or cooling so that to some extent both thermosetting and thermoplastic systems can be compounded. Material is fed into the mixer, which might have a capacity of say, 500 kg, through a chute which can be closed by a hydraulic ram, thus allowing pressure to be applied to the mixing process, thereby reducing cycle time. The product of the mixing is not appropriate for direct introduction to the mould, being lumpy in texture, and so it is usual to feed it first of all to a mill from which it emerges as a sheet which can be granulated, or alternatively to an extruder from which it comes in a suitable form for pelletising. Both secondary processes will of course give rise to further mixing. For reasons of speed, output, and economy, internal mixing is also used with elastomers as well as plastics, although here roll mills are also used (Table 5.1).

Mills also find wide application in polymer compounding, roll mills being the most important. These come in different sizes and complexity

TABLE 5.1
COMPOUNDING METHODS FOR POLYMER COMPOSITES

Mixing	
Tumble mixing	Extrusion
Paddle, Z and Σ blade mixing	(single, twin screw)
Ribbon blending	Laminating
Air/fluid mixing	Pultrusion
Solution, emulsion blending	Filament winding
Internal mixing	Spray mixing
Mills	
Roll mills (2, 3, etc.)	
Colloid, disc, pin mills	

of up to five rolls with facility for temperature control. A two-roll mill has parallel rollers rotating in opposite directions and at different speeds, with typically one roller being at a slightly different temperature to the other. The distance between the rollers (nip) can be altered to effect shearing of the polymer and filler mixture added to the nip. Although labour-intensive, this can be an efficient way of mixing but rate of output is low.

The third important method of compounding is through use of an extruder. A screw extruder by its very nature is not only an equipment for fabrication, but is also a mixer, and it may in fact be specially designed for the latter purpose, with both filler or masterbatch and polymer being fed to the hopper. In a single-screw extruder the mixing is one of large-scale convective movement coupled with localised shear and molecular diffusion, requiring, with the help of temperature control, an optimisation of residence time and viscosity, the amount of mixing being determined by the product of shear rate and time. Thus a great deal of design is involved in deciding the nature of the channel depth, angle of screw helix, pressure, mixing head, and speed. Extruders designed especially for compounding include mixed-feed, hot-melt, and cold-feed extruders. An extruder which has found application in elastomer compounding is the Transfermix, having convoluted profile instead of the more conventional regular cylinder type. Twin-screw extruders, although more complex and expensive, have two extruders side by side and have several advantages over single-screw machines, including low wear, more versatility as regards feed, self-cleaning ability, and better homogenizing power. In these, the screws may rotate in the same or opposite directions.

All three methods have the disadvantage with fibrous fillers that some breakdown of fibre occurs, and although this may be reduced by taking care, this, on the other hand, may be at the expense of other desirable factors such as wetting, voiding, and efficiency of mixing. Where retention of fibre length is important, use can be made of a coating process in which a number of strands of continuous fibre are die-coated with a thermoplastic melt and then pelletised after cooling. The apparent disadvantage of this approach in producing fibre-agglomeration is overcome to some extent by the subsequent mixing prior to injection moulding. Whatever the method chosen, the usual practice nowadays for injection moulding of fibre composites is via pellets. Although not much used in this way, precompounded pellets are capable of being used in compression moulding, but no further improvement in mixing will be forthcoming in this procedure.

Fabrication

Having discussed compounding we can now turn to fabrication methods, remembering that in, say, extrusion and a number of processes to be described later for thermosetting resins such as polyesters and epoxies, compounding and fabrication may overlap. Many of the techniques for processing polymer composites are the same as for the unfilled polymers, but require some modification, possibly because of the higher viscosities which need higher moulding pressures to give the same output. Some processing techniques, however, are specific to the fabrication of composites or to one particular class of composite, because of some special requirement. Some indication of both classes is given in Table 5.2. The

TABLE 5.2
COMPOSITE POLYMER MOULDING TECHNIQUES

General methods	Polyester/epoxy resin methods
Open casting	Hand lay-up
Slush casting	Spray lay-up
Powder coating	Thermoforming (vacuum/pressure
Hot compression moulding	bag, matched metal moulding)
Cold compression moulding	Foam reservoir moulding
Transfer moulding	Resin injection/transfer
	moulding
Injection moulding	Vacuum impregnation
Extrusion	Filament winding
Calendering	Centrifugal moulding
	Continuous sheet moulding
	Pultrusion

principle of processing of composites, like that of the parent polymer, is to arrange for the material to be a fluid which can then be formed by means of solvent, unpolymerised reactants, heat or pressure, and combinations of the same. Removal of the forming agent will allow the material to assume a less fluid, fabricated state. The effect of the filler is to stiffen the system and the role of the forming agent is to overcome this. Generally, the use of pressure within a restricted temperature rise is the preferred approach, except for special cases of very high viscosity systems, or where the composite is prepared in a thin film.

Compression moulding, which is as old as the history of composite materials, being used for the fabrication of phenolic resins, is therefore the oldest forming technique. In this, the composite in powder form is poured into the cavity of a mould. It softens on heating, and pressure is then applied to fill the mould. Viscosity first decreases and then increases as curing proceeds. In a typical moulding of a phenolic, aminoplast or silicone composite, pressures in the range 20–60 MN/m^2 are applied for a period of 1–2 minutes. The mould is then cooled to a temperature such that the shape rigidity of the moulded object is retained on ejection from the mould. The technique has the disadvantage that some material is lost as flash, since the mould must be overfilled to allow pressure to build-up; it is also slow compared with, say, injection moulding. On the other hand it is clearly useful for short runs which would not really justify the use of injection moulding. Another process which has been used for many years is cold-moulding, in which the composite is preformed under high pressure and then cured in an oven or other heated apparatus, including normal compression moulding equipment. In transfer moulding, the composite powder is heated in a chamber outside the closed mould and then forced by pressure through channels and a gate into the mould and cured as before. Because of the flow some orientation of molecules and possibly filler can take place, but on the other hand because the mould is closed flash is eliminated. Also more delicate inserts perhaps connected with the preparation of macrocomposites, can be made by this technique, and there is less interference of volatiles on moulding quality.

A process which again involves moulding of thermosetting polymer composites, especially for large regularly shaped structures, is lamination, which as the name implies is the building up of sheet structures which may be flat, tubular or rod-like, etc., depending upon the shape of any former which may be used. The overall process takes place in three stages, impregnation, drying, and finally forming by curing. In the process, paper or woven (or non-woven) cloth is impregnated with resin or resin solution, a doctor blade or squeegee being used to control uptake. A drying-oven drives off solvent and advances cure, after which the sheets are cut if required, and stored in low-humidity rooms. To make a laminate a number of sheets are laid on top of each other, the top one possibly having a design, such as a printed wood pattern if the laminate is to be used, say, for a simulated wood tray or table top. They are then placed between platens, which themselves may be chosen to provide a polished, matte or embossed finish, and cured by heat under pressure. For a phenolic-based system the overall cycle time for this may be of the order of ninety minutes. For the preparation of tubes, an impregnated paper system may be passed over a heated roll and wound on to a steel mandrel which is placed in a special oven for pressure curing.

A very important fabrication technique also used for the production of long, simple, and regular geometries, but this time for thermoplastic composite systems, is extrusion. As mentioned earlier the modern extruder is fed by a screw, the best extrusion conditions being obtained by matching screw characteristics to a particular system. In addition the heating programme for the barrel of the extruder must take into account the shear heating which itself will depend upon screw speed and the non-Newtonian behaviour of the composite. Despite the extra cost, twin-screw extruders have advantages over single-screw extruders in that they are more versatile as regards screw modification. They can also accommodate bulkier feeds and there is less machine wear.

As far as overall importance is concerned, the most used fabrication process, certainly with reinforced thermoplastic composites, is injection moulding, the principle of which can be seen in Fig. 5.6. Granules of

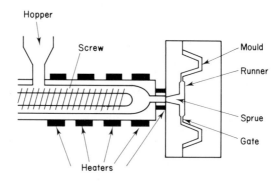

FIG. 5.6. Injection moulding machine.

precompounded polymer and filler are fed by means of a hopper to a heated cylinder where the material is softened but not to such an extent that demixing occurs. This material is driven forward, usually by a screw (although earlier equipment used a ram) which also acts as a metering device as well as completing mixing, through a nozzle and into the mould cavity. The material is allowed to cool, the mould opens and the article is freed by ejector pins. The equipment, which allows moulding to take place within seconds and so for long runs is greatly preferred to compression moulding, was originally developed for thermoplastics, but it is now used on thermosetting polymers and foamed polymers. For reinforced polymers, higher injection pressures are required, putting up the already high costs of the equipment, but cycle time is shorter, partly because of the shorter cooling times arising from increased heat transfer and the stiffening action of the filler, thus doing away with the need for much cooling before ejection. Likewise it is possible to mould thinner sections, because of the reinforcing effect. For fibrous fillers there is a tendency for orientation of fibre which can be restrained to some extent by appropriate attention to gate size and position. The other aspect which arises with the use of hard fillers, is the possibility of wear, but this also can be minimised by correct moulding procedure and of course, by the use of more resistant steels. High temperatures and excessive work can cause separation, say of glass fibres from the matrix, but large sprues and runners both cut down this tendency and also reduce glass fibre breakdown. They also permit high rates of flow.

Injection moulding of foams is possible with or without fibre reinforcement, an advantage here being that much lower moulding pressures and therefore less costly machinery is needed. Cycle times are longer but moulding stresses are lessened. Another development in injection moulding, quite apart from microprocessor-based robot operation, is in its application, as indicated above, to thermosetting systems. The interest in this came about because of the plasticising effect of the screw. Barrel temperatures are different and lower than for thermoplastics but the technique is capable of moulding phenolic and melamine-type composites, and also glass-reinforced polyesters.

In many ways the most exciting development in injection moulding in recent years is reaction injection moulding (RIM) and for reinforced polymers, reinforced RIM (RRIM). It is designed, as the name suggests, for reacting systems, particularly polyurethanes, but also to a more limited extent, for other polymers like epoxy resins and polyesters where no appreciable volatile products are formed during cure. It is a chemical and moulding operation combined into one, with the chemical generally being catalysed for fast reaction, leading to cycle times of no more than a few minutes. The ingredients including filler are synchronously metered into a mixing head with viscosity, temperature, and pressure being controlled to better than 1 per cent. Mixing is through turbulence and after each injection, the mixing chamber is cleaned by means of a ram. Pressures are less than in conventional injection moulding and for reinforced polymers, there are several advantages over other moulding methods such as sheet and bulk moulding compound techniques to be described later. Although other fillers such as mica flake and carbon fibres have been evaluated, the most common fibre used in RRIM is glass, with some competition between hammer-milled and chopped fibre, the former possibly needing for equivalence a greater concentration of filler but with a substantial proportion of fibres of less than critical length, whilst the latter is specially made to be of shorter than usual length (Fig. 5.7).[19] Some orientation of fibre occurs as in other injection

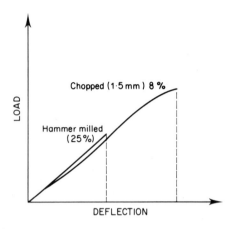

FIG. 5.7. Impact behaviour of hammer-milled and chopped glass fibre reinforced polyurethane (RRIM). (Note: area under curve is a measure of energy to failure.)

moulding processes. The present great interest in the technique is in connection with automobile applications, especially in microcellular polyurethane, but there is also interest in the production of furniture and household appliances. Compared with other injection moulding the method is less expensive because of the lower clamping pressures for the mould, much of the cost of an injection moulding machine being in the ancillary components for producing high pressure. In addition, the technique is energetically favourable.

As well as the above-mentioned processes for the moulding of polymer composites, there are many others which will not be discussed here, but there is one very large and important group of methods which in view of their position in the field of reinforced plastics deserve a special mention. These are the techniques which have been closely identified with unsaturated polyesters, but are used in other curing systems. Further, whereas all the methods described so far are matrix directed in the sense that they are primarily designed for processing of polymers themselves, the filler merely introducing a modification, the methods to be considered now are composite-directed in the sense that they are specifically designed for integrated moulding of composites, with long and continuous fibres as well as shorter fibres dominating choice of technique. Although in some of the fabrication processes the filler and resin are brought together at the same time, in others use is made of compounded materials and it is useful to outline the nature of these before describing actual processes.

Being compounded implies that they are essentially stable until heated and indeed they are used in heat curing situations. The first of the compounded materials are the dough moulding compounds (DMCs) and bulk moulding compounds (BMCs), the former generally indicating an isophthalic polyester-based resin and the latter an orthophthalic polyester, although the names may be used interchangeably. Both may contain glass or other fibres with DMCs usually containing between 15 and 20 per cent glass fibre appropriate to the viscosity needed for fabrication, whilst the BMCs generally have a lower glass fibre content but contain up to 20 per cent particulate filler such as calcium carbonate, talc or clay to provide a dry handle, reduce cost, and of course, reduce shrinkage on cure. Mixing may be carried out in a Z-blade mixer with the compounds being extruded as rope. Sheet moulding compound (SMC), and its thicker version (TMC), approximately 5 cm compared with just over 0·5 cm, are similar to the above but with a higher fibre concentration of 20–35 per cent and longer fibres of about 2–5·5 cm, which may be glass, carbon or hybrids of the two, etc. They also contain thickening agents such as calcium or magnesium oxide and a particulate filler, as for BMCs, and are prepared between polyethylene, i.e. low adhering, continuous film, passed through rollers and wound on to a roller ready for use.[20]

Prepregs are impregnated sheets, tapes or woven fabrics or random and unidirectional fibre reinforcement, such as glass, carbon, aramid or epoxy resin, but not containing other major filler. They are made in a similar way to SMCs or with solutions of resin, the solvent being recovered by evaporation after impregnation of the cloth, roving, etc. Preforms find application where an even distribution of fibre is required but the moulded shape may be complicated. They are prepared by building up a structure by spraying fibre on to a wire mesh screen shaped to match the final mould. A binder to hold the fibres is next sprayed on to stabilise the shape. The binder is chosen to be styrene-insoluble, so that the preform does not break down on application of the polyester resin system.

Turning now to ways in which fabrication of reinforced plastic can be carried out, it should be pointed out that these may be classified as hot and cold moulding processes, or as hand and automated (including semi-automated) processes. The latter approach is the basis of the following description, as it emphasises that hand moulding methods are still very important.

In hand lay-up techniques as used in boat building, the mould is first treated with a release agent. On to this a resin-rich thixotropic gel coat is applied. This will be the surface of the final object, and so has important aesthetic requirements. Once this is tacky, thereby providing a key for the next stage, a tissue or CSM layer is applied, the latter being followed by successive layers of woven fabric, etc., as alternate layers. When the desired thickness is reached a surface tissue finish or possibly a pigmented resin is applied to complete the finish. Following the gel coat stage, the build-up of the reinforced plastic may also be carried out by a spray lay-up process in which glass fibre and resin at about 30 per cent relative concentration is sprayed from a spray gun. As before, alternate layers of woven roving, etc., may be used. Despite some critical demands on the skill of the operator, the spray lay-up method tends to be faster overall than hand lay-up. The two processes find use for the preparation of large objects (as well as small), and where small production runs are needed.

Turning to automated and semi-automated techniques, let us first have a look at hot press or matched metal moulding. These offer better moulding control, more rapid moulding, and better surface finish. The method uses preforms, CSM, SMC, prepreg, DMC, and continuous filament in the form of swirl mat. Where appropriate catalysed resin is added to the shaped filler in the mould. In other cases such as prepreg or SMC the resin is already present. The composite is cured under heat and pressure to give moulding times of a few minutes, i.e. much less than what might be days in the case of hand lay-up methods. Because of the higher moulding temperature the technique can accommodate higher concentrations of fibre.

Of other techniques, mention is made of vacuum bag moulding of wet lay-up, spray-up, and prepreg systems in which the composite is drawn to the shape of the mould by means of a vacuum applied to a flexible bag covering the composite layer, before moulding is carried out. Alternatively pressure may be applied outside the flexible bag when we have pressure bag moulding. As in principle, the differential pressure can be greater than in vacuum moulding, the equipment is correspondingly more robust. If instead of having one rigid side and one flexible side as in the above, two rigid faces are used, then resin and fibre composite can be forced under pressure by a resin injection moulding approach to fill the mould after the faces have been coated with gel coats. Resin injection.

which has had a chequered history, is claimed to have a number of advantages including versatility. Again vacuum can be applied to provide the driving force for filling the mould, instead of direct pressure. Of other methods closely associated with short fibre processing, there are, in addition, transfer moulding and straightforward injection moulding. Casting by centrifugal moulding of cylindrical items is another important technique, with compacting being achieved for a variety of compounded systems, by rotation of the mould thus providing the centrifugal driving force.

Two fabrication methods particularly relevant to continuous fibre systems are filament winding and pultrusion. In filament winding, fibre is fed under tension through a resin bath and wound on to a mandrel. Successive layers can be wound in different patterns to reduce anisotropy in final properties. After cure the moulded object, typically tube or tank, can be removed for use. In this technique in addition to single filament unidirectional filler, prepreg tape can be used. Filler content in this process is usually high, up to 80 per cent by weight, so that high stiffness and strength, in the reinforcement direction, are possible. The object may be finished with a gel coat. Pultrusion is a continuous process with production rates of about $1 \, \mathrm{m/min}$ involving oriented glass or perhaps carbon fibre, which is fed through a liquid resin bath and through a heated die, or alternatively the resin may be injected into the heated die. Glass fibre contents up to 70 per cent are common so that lower tensile properties than for filament winding may result, although on the other hand transverse properties may be better. The technique can be used to provide a wide range of profiles or to provide prepreg stock for use in other fabrication methods.

Summarising, we have seen that a wide range of fabrication techniques are available for the production of reinforced and filled plastics and elastomers. Which technique may be chosen in practice will depend upon a number of factors including, property requirements, cost, shape, and size of article, with convenience also playing a part.

REFERENCES

1. OSTWALD, W., *Kolloidzschr.*, **38**, 261 (1926).
2. FOX, T. G., GRATCH, S. and LOSHACK, S., in Eirich, F. R., (Ed.) *Rheology*, vol. 1. Chapter 2, Academic Press, New York (1956).
3. WILLIAMS, M. L., LANDEL, R. F. and FERRY, J. D., *J. Am. Chem. Soc.*, **77**, 3701 (1955).

4. ANDRADE, E. N. DA C., *Nature Lond.*, **125**, 309, 582 (1930).
5. WEISSENBERG, K., *Nature Lond.*, **159**, 310 (1947).
6. EINSTEIN, A., *Ann. Physik*, **19**, 289 (1906); **34**, 591 (1914).
7. GUTH, E. and SIMHA, R., *Kolloidzschr.*, **74**, 266 (1936).
8. CROSS, N. M., J. *Coll. Interface Sci.*, **33**, 30 (1970).
9. See COGSWELL, F. N., *Polymer Melt Rheology*, Plastics and Rubber Institute, London (1981).
10. MOONEY, M., J. *Coll. Sci.*, **6**, 162 (1951).
11. See, e.g., KATZ, H. S. and MILEWSKI, J. V., *Handbook of Fillers and Reinforcements for Plastics*, p. 30, Van Nostrand Reinhold, New York (1978).
12. HAN, C. D., *Rheology in Polymer Processing*, p. 182, Academic Press, New York (1976).
13. See SHERMAN, P., *Industrial Rheology*, Academic Press, London (1970).
14. SIMENEV, Y. and IVANOV, Y. *7th Int. Congress of Rheology*, p. 384, Swedish Soc. of Rheology, (1976).
15. PLOSHACK, A. P., in Paul, D. R. and Newman, S. (Eds), *Polymer Blends*, vol. 2, p. 340, Academic Press, New York (1978).
16. LEE, B-L and WHITE, J. L., *Trans. Soc. Rheol.*, **19**, 481 (1975).
17. CASTOR, W. S. and MANASSO, J. A., in Seymour, R. B. (Ed), *Additives for Plastics*, vol. 1, Academic Press, New York (1978).
18. BOONSTRA, B. B., in Blow, C. M., *Rubber Technology and Manufacture*, p. 247, Butterworths, London (1971).
19. CHISNALL, B. C. and THORPE, D., *Proc. 35th Ann. Tech. Conf.*, RP/C Inst. S.P.I. (1980).
20. WEATHERHEAD, R. G., *FRP Technology*, Applied Science Publishers, London (1980).

Chapter 6

POLYMER BLENDS AND BLOCK COPOLYMERS

INTRODUCTION

Although mention has already been made of polymer blends based upon mixtures of two or more polymers, their properties are sufficiently distinct from those of other composite polymeric systems to merit separate consideration. To a large extent this distinction arises not only from variation in relative concentration of materials which are not so dissimilar in properties as they are in conventional composites, but also from differences in chemical nature and from differences in viscosities at the time of fabrication, reflecting the common deformability of the separate components. The chemical interaction can mean, depending also on effort expended on mixing, materials which span the scale from completely homogeneous mixtures on the one hand to non-interacting two (or more) phases on the other. Further, since block copolymers and the related graft copolymers can also form two-phase systems which although interconnected by primary bonds, have properties akin to polyblends and in a structural sense resemble polymer composites, they too are included. The identity is reinforced by the interchangeability of block copolymers and polymer blends in many applications. They perhaps differ from polyblends in that their science and technology have developed apace whilst it is only in recent years that the science of polymer blends is just beginning to be understood.

POLYMER BLENDS

Like other composite systems, polymer blends, or polymer alloys as they are sometimes known, tend to have properties intermediate between those of the separate components, but in addition they frequently have properties characteristic of the blend as a whole. They have been known for many years, first coming into prominence as materials of high impact resistance such as high-impact-resistant polystyrene (HIPS) which was made from a blend of styrene butadiene rubber and polystyrene as long ago as 1927, and later as acrylonitrile–butadiene–styrene high-impact resins (ABS) in 1952. They are establishing a new importance as secondary or recycled polymer systems made from plastics waste, and as systems which lead to biodegradable composites which find use in 'throw-away' plastic carrier bags. In the case of conventional polymer composites miscibility between components is not usually of concern, indeed the phase separation is crucial to the high modulus and strength properties, provided there is a facility for stress to traverse both phases. But for polymer blends compatibility is of prime concern in explaining the properties of these materials and the related copolymers. For example, a polymer called on to act as a plasticiser for another polymer should be miscible at a molecular level to be useful, whilst an elastomeric filler to provide good impact resistance to a plastic must have some degree of incompatibility to be able to dissipate stress energy whilst retaining the stiffness of the parent polymer. Mixing and compatibility have not necessarily the same connotation although the ultimate in compatibility will in principle lead to intimate mixing, although kinetic restrictions associated with long-chain polymers may mitigate against achieving this state, at least in a reasonable time. Thus even polymers of the same type, which would be considered to be compatible, might find some difficulty in interspersing with each other. It is useful at this stage to examine the meaning of miscibility and to see to what extent partial or complete immiscibility influences phase separation.

As in the case of any equilibrium process, whether chemical or physical, in order that spontaneous mixing of different components might occur, the free energy of the mixed system must be less than that of the sum of the free energies of the separate components under a given set of experimental conditions. Or mathematically:

$$\Delta G_{\text{mix}} = \Delta H_{\text{mix}} - T\Delta S_{\text{mix}} \leqq 0 \tag{6.1}$$

where ΔG_{mix}, ΔH_{mix}, and ΔS_{mix} are the free energy, enthalpy, and entropy changes which take place on mixing, at temperature T. Schematic representation of the various changes in free energy which can occur, expressed as a function of concentration, since the above are extensive changes, is presented in Fig. 6.1. In the first case, (a), the total free energy

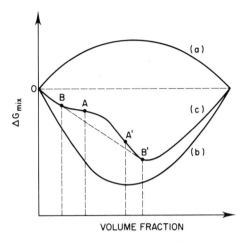

FIG. 6.1. Change of free energy with concentration.

would be positive at all concentrations and such components are thermo-dynamically, but not necessarily kinetically, immiscible. Curve (b) takes on negative values for the free energy change and so such a system is miscible in all proportions, at least in principle. Curve (c) although of negative free energy, exhibits points of inflexion at A and A'. As a consequence, a mixture of composition between these two composition limits (spinodals) is metastable and can achieve a lower free energy state by separating into two phases of composition given by B and B' (binodals). As might be expected the positions with respect to composition at which spinodals and binodals occur for a given system, if they occur at all, will depend upon other thermodynamic parameters such as temperature, and behaviour as illustrated in Fig. 6.2. The minimum occurs at what is called the lower critical solution temperature (LCST) and the maximum at the upper critical solution temperature (UCST). This also means that systems which may be compatible at one concentration and temperature may give rise to phase separation on changing the temperature. Although examples of both kinds of behaviour are well known in non-polymeric binary systems, polymers tend to form LCST systems only, examples of these being polystyrene/polyvinyl methyl ether and polyvinyl chloride/ethylene–vinyl acetate copolymers. Once it is understood how miscibility and partial miscibility can arise, it would be of interest to scientists and technologists to be able to predict the behaviour of pairs of given polymers *ab initio*, and so it is instructive

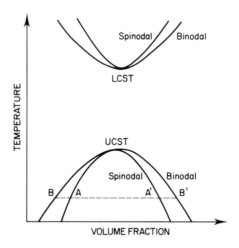

FIG. 6.2. Influence of temperature on the phase behaviour of two-component systems.

to have a look at the factors affecting polymer miscibility at the molecular level.

The starting point for this is again eqn. (6.1) and involves an approach pioneered by Flory.[1] In this ΔG_{mix} is not determined directly but separately through estimations of ΔS_{mix} and ΔH_{mix}. The conclusions are limited but at least highlight important influences. ΔS_{mix} will be composed of a probability term relating to the number of ways of arranging the polymer monomer units for each species on a hypothetical lattice, thereby constituting configurational entropy. In addition there is a term relating to the specific ordering effect of one type of monomer unit on the other, an interaction entropy, but this is usually ignored in subsequent calculations. From the former contribution we get:

$$\Delta S_{mix} = -R(N_1 \ln X_1 + N_2 \ln X_2) \qquad (6.2)$$

where N_1 and N_2 are the mole fractions of the separate components and X_1 and X_2, the volume fractions. For polymers having connecting bonds between individual lattice points, the number of arrangements is less than for free units, such as might be represented by, say, solvent molecules. Thus the entropy of mixing is less than for simple molecules and approaches zero at high molecular weight. This overall small positive term combines with temperature to give a small $-T\Delta S_{mix}$ term. This means that, within limitations of the theory, ΔH_{mix} must be approximately zero or negative for ΔG_{mix} to adopt a negative value, a first prerequisite for thermodynamically spontaneous mixing, as mentioned earlier. It has been shown that for systems involving dispersion force interaction only, the following equation can be derived:

$$\Delta H_{mix} = V(\delta_1 - \delta_2)^2 X_1 X_2 \qquad (6.3)$$

where V is the volume of the mixture and δ_1 and δ_2 are the solubility parameters of the two components. (The solubility parameter is defined as $\delta_L = (\Delta H_L - RT)^{\frac{1}{2}}/V_L^{\frac{1}{2}}$ where ΔH_L and V_L are the latent heat of vaporisation and molar volume of a liquid species. For polymers, being non-volatile, tables have been constructed for calculating the solubility parameter. However, as might be expected, the values are near to the values determined experimentally for the liquid monomers.) As will be realised from eqn. (6.3), for systems associated with dispersion forces only, the value of ΔH_{mix} will be either positive or zero. Combining the equations for enthalpy and entropy we have:

$$\Delta G_{mix} = V(\delta_1 - \delta_2)^2 X_1 X_2 + RT(N_1 \ln X_1 + N_2 \ln X_2) \qquad (6.4)$$

This equation implies that mixing of two such polymers is not very favourable. Going outside the above enthalpic treatment to consider exothermic mixtures arising from dipolar and other strong interaction for example, at first sight the negative values of enthalpy suggest an approach to compatible polymer systems. However, it is in these cases that the interaction entropy term becomes significant and so any indications from theory may not be as fruitful as might be hoped. Clearly some compromise may be needed. During the last decade or so considerable effort has been expended in an attempt to find polymeric systems which, within the limitations of deciding what is compatible, are believed to form true mixtures. Some of these are shown in Table 6.1, but as is clear, these are in a minority, compared with the total number of possible polymer pairs. As pointed out by Bucknall,[2] of over three hundred pairs of polymers considered by Krause,[3] only about a tenth of these mix to form a single phase and in many cases where miscibility is apparently found, this is usually over a limited concentration range. It is interesting to note that the pairs which are reported to be compatible are polymers which can enter into specific secondary bond interaction such as dipole–dipole interaction. This serves to emphasise the dominating effect of a negative enthalpy of mixing.

Although, as mentioned, some attempt has been made to recognise miscible systems, in fact the actual assessment is not at all easy. Typical

TABLE 6.1

EXAMPLES OF COMPATIBLE POLYMER BLENDS

Polymer 1	Polymer 2	Compatibility range (vol per cent of polymer 1)
cis 1,4 Polybutadiene	Poly(butadiene-co-styrene) (75/25)	20–80
Polyisoprene	Poly(butadiene-co-styrene) (75/25)	50
Polymethyl styrene	Poly 2, 6-dimethyl-1, 4-phenylene ether	0–100
Polyacrylic acid	Polyethylene oxide	> 50
Nitrocellulose	Polyvinyl acetate	0–100
Polyisopropyl acrylate	Polyisopropyl methacrylate	0–100
Polyvinyl acetate	Polymethyl acrylate	50
Polymethyl methacrylate (iso)	Polymethyl methacrylate (syndio)	0–100
Polymethyl methacrylate	Polyvinyl fluoride	> 65
Polyethyl methacrylate	Polyvinyl fluoride	> 49
Polyvinyl acetate	Polyvinyl nitrate	0–100
Polyvinyl acetate	Poly ε-caprolactone	> 49
Polyvinyl chloride	Poly α-methyl styrene/methacrylonitrile/ ethyl acrylate (58/40/2)	0–100
Polyvinyl chloride	Poly ε-caprolactone	> 49
Polymethyl methacrylate	Polyvinylidene fluoride	> 65
Nitrocellulose	Polymethyl acrylate	0–100

of the criteria which have been used are optical clarity, the moving together of the separate glass-transition temperatures, a corresponding 'fusion' of loss peaks in dynamic mechanical studies, and the appearance of a LCST on heating a mixed polymer system. In the case of optical clarity, complications can arise if the two polymers have the same refractive index, although change of temperature should help to discriminate as it is unlikely that the two polymers will have the same temperature coefficient of refractive index also. Further, even if the polymers are not mixed at a molecular level, but the particle sizes, indicating mode of preparation, are less than the wavelength of light, then again the composite will appear to be homogeneous, although actually not. Confirmation in this situation might be obtained from electron microscopy which can resolve particle sizes less than those possible by optical microscopy. An unequivocal assessment might be expected from the appearance of a single glass-transition temperature for systems where the two polymers have normally separate and distinct glass-transition temperatures. However, at least one case has been reported in which this is not necessarily true, a common solution of polymethyl methacrylate and polyvinyl acetate when freeze dried producing a polymer blend which on first heating gives only one glass-transition temperature, reflecting the intimate mixing provided by the solvent, but which on reheating shows two transition temperatures, demixing having occurred as soon as one of the polymers becomes mobile.[4] Dynamic mechanical analysis not only suffers from the same disadvantage, but has a further problem in that to reveal the two separate glass transitions of a two-phase system there must be mechanical coherence between the two. If there is not, then the technique may only exhibit the relaxation of the continuous phase. Some indication of whether this is happening may be obtained from the position of the loss peak, or peaks, relative to one or other of the loss peaks of the individual polymers. Even the LCST approach must be carried out with care in that kinetic effects control the rate at which equilibrium can be re-established at the higher temperatures, and also for the reason mentioned in connection with the differential refractive index temperature coefficient. Other techniques which have also been used involve the observation of common solvent clarity, and the mechanical nature of a film prepared from solution, a brittle crumbly film indicating incompatibility.

For many applicational purposes, true miscibility is not the main prerequisite for useful properties. Indeed, for some properties phase separation is an advantage in that although for some applications the primary matrix properties are what are required, in others where mechanical shock is to be expected, impact resistance provided by a soft dispersed filler is of importance. In all cases, however, some degree of mechanical coherence is required, in order to transfer stress across the interface between the two phases. In some cases, this is achieved through frictional forces as in the example where the composite has been cooled after fabrication, and where the internal phase has shrunk less than the continuous matrix, so that the former is under pressure from the latter. In this way physical contact is maintained and so the composite is capable of transmitting any mechanical forces across the interface. In other cases, this continuity has to be achieved by what might be called compatibilising techniques. This does not mean complete compatibilisation to provide an intimate mixture of the two polymers, but at least to be able to produce a surface compatibility in the same way that polyethylene, for example might be surface oxidised to promote compatibility for wetting, etc., as reported in Chapter 1. Essentially, attempts to provide this compatibilisation are attempts to provide strong interaction across the interface and perhaps at the same time to ensure that this is sufficiently strong to resist fracture on extensive deformation or that it provides some flexibility to accommodate any different micromechanical response of the two components. The main 'thrust' of research has been to use graft and block copolymers to promote compatibility, random copolymers having little influence. The mode of operation of the block and graft copolymer is believed to be through a molecular interaction between each block and the corresponding homopolymer.

Other attempts have been through the use of mutual plasticisers, compatible fillers, and direct chemical cross-linking between the two components.[5] The latter can be done either by incorporation of a mutually reactive chemical in the blend compounding, followed by reaction, or possibly through mechanical working and degradation of the two, when free radical intermediates might react to form, again, block and graft copolymers. Generally speaking block copolymers rather than graft are preferred as, say, in the compatibilisation of polybutadiene and polystyrene with block copolymers of the two. The improved compatibilisation also increases the amount of the two components which can be mixed, a graft polystyrene/polybutadiene being able to increase the amount of polybutadiene homopolymer from 10–40 per cent and still produce a useful blend with polystyrene. The use of copolymer compatibilising agents also increases the solubility in mutual solvents. The length of the blocks is of importance, high molecular weight being less effective than block lengths of about 10–15 monomer units, which incidentally is not so very different from the basic average flow unit of molten polymers, suggesting a kinetic miscibility influence for practical compounding. The presumed surface activity role of a block or graft copolymer is shown in Fig. 6.3.

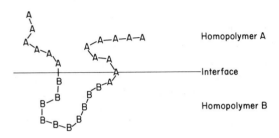

Fig. 6.3. Surface-active role of a block copolymer.

Mention has already been made of the influence that the actual preparation method might have on the miscibility, apparent miscibility, and compatibility of polymers in blend preparation. It is appropriate now, therefore, to consider ways by which this may be carried out. The methods include: melt blending, solution blending, latex blending, and in-situ polymerisation (possibly involving some incidental block copolymerisation). For the manufacture of the important class of toughened polyblend polymers, emulsion and the related suspension polymerisation methods have largely replaced direct blending, but this still finds application at the commercial level by the melt blending of a rubber and thermoplastic polymer using the conventional techniques of extrusion, internal mixing, and milling. For example, the techniques are used in the toughening of polyvinyl chloride with synthetic elastomers such as nitrile rubber, chlorinated polyethylene, etc., and of polypropylene with ethylene–propylene rubber, etc. Masterbatching of rubber and grafted thermoplastic is also carried out in this way. The action of shear can cause chain scission and the possible formation of graft and block copolymers which in turn have an effect in dispersing one of the polymers in the other. The nature of a polyblend dispersion prepared in this way will depend upon the period of mixing, temperature, shear field, and type of equipment, and of course, on the rheological properties of the component polymers.

Solution blending is not of importance for bulk production of polyblends in view of the problems of disposal of solvent, but it is used with surface coatings, though even here there is always the possibility of precipitation during evaporation. The latter is acceptable if the mixture is going to be processed. An interesting feature of the preparation of polyblends from solution is that, in principle, providing the relative concentrations of the two polymers are not too different, if a particular mutual solvent is better in a thermodynamic sense for one polymer than the other, the chains of the former will be more extended in space than the other, and so will tend to form the continuous phase on evaporation. Thus by appropriate choice of solvent, determined largely by matching of solubility parameter, it is possible to get structural inversion (see Fig. 2.1). In this way, for example, it is possible if one polymer is a rubber and the other a plastic, to obtain either a reinforced plastic or reinforced elastomer merely by choice of casting solvent.

Latex blending is a convenient way of mixing two different polymer latexes, provided the emulsifying agents are of the same type so that precipitation does not occur. It is a particularly convenient technique in the light of the low viscosities of the supporting liquid, usually water. The presence of stabiliser can be a problem for applicational purposes, and also the blend may need further heat treatment for consolidation in product fabrication.

The direct preparation of a polyblend from one, at least of the monomers if not both, is of considerable technological importance. The various forms include interpenetrating polymer networks (IPNs), in which one of the polymers is swollen by the monomer of the other following which polymerisation of the latter is carried out, and simultaneous interpenetrating networks (SINs), in which both monomers are polymerised together. A related system, interpenetrating elastomeric network (IEN) involves coagulation of a mixed elastomeric dispersion in the presence of a cross-linking agent which completes the network formation. In addition there are semi-IPNs, gradient IPNs, and interstitial composites.[6] In a typical preparation of high-impact polystyrene (HIPS) a solution of butadiene rubber in excess styrene is heated with a peroxide catalyst. In the early stages polystyrene is in the minor composition, so that the composite is essentially one of polystyrene dispersed in a rubber solution, but as polymerisation proceeds the polystyrene begins to dominate the system and phase inversion occurs, forming as desired a polystyrene system reinforced by polybutadiene. A number of factors will affect the final structure including the possibility of any grafting during polymerisation, stirrer speed, temperature, and the nature of the original elastomer. In one of the different ways in which ABS terpolymer blends are made, a polybutadiene rubber latex is first prepared by emulsion polymerisation. To this is added styrene and acrylonitrile monomers in the presence of more water, soap, initiator, and transfer agent. After addition of antioxidant following polymerisation, the system is coagulated, washed, and dried.

Before turning to the properties and applications of polymer blends, since morphology of the composites must have a bearing upon properties, it is relevant to discuss some of the factors which affect morphology and to examine to what extent they may be controlled. As we have seen, compatibility, certainly over wide composition and temperature conditions, is the exception, and therefore we would expect most polymer blends to show phase heterogeneity. That this is the case is well substantiated by a number of direct observations, as well as by indirect implication from a number of experimental techniques. The techniques used in morphological studies include optical and electron microscopy, gel separation, differential thermal analysis and differential scanning calorimetry, dynamic mechanical analysis, etc. It is found that there are three principal types of morphology. These are (a) one continuous phase throughout which is dispersed the second phase, similar to the conventional filled polymer, (b) two continuous phases and (c) two discontinuous phases. The actual phases may be separate polymers or individual mixed compositions, or they may even be regions, or domains as they are commonly termed, within which there are microdomains of separate compositions (Fig. 6.4). Considering the first class and to some extent the third class, the domains will differ in size, shape, and number depending upon such factors as the presence of compatibilising agents, including copolymers, and the method of preparation and fabrication. For example, in the case of normal HIPS domain sizes of 1–10 microns are typical whereas in an ABS polymer system, the domains might well be a tenth of this in size. In some systems which might be expected to constitute the third class, i.e. blends prepared from latex mixtures, it has been shown that domains much larger than would be expected from the dimensions of the individual latex particles, are found. Electron microscopy shows that one component, the rubbery one, has in fact coalesced and is thus to some extent, continuous. A considerable amount of work has gone into the study of the morphology of IPNs and it is now fairly clear that the factors which control the morphology of such systems are compatibility, interfacial tension between phases, cross-link density of

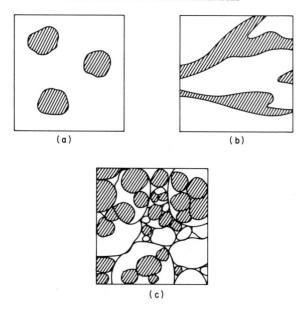

FIG. 6.4. Phase separation in polymer blends (see text).

networks, composition, and polymerisation method. It is found that the
original polymer, i.e. the one swollen by monomer, tends to form the
continuous phase with the second polymer showing less phase continuity. Stirring during polymerisation can produce phase inversion,
especially if the original polymer is in the lower concentration. For melt
blended systems, the factors which control continuity are concentration
and relative viscosity. The influence of fabrication conditions on morphology is further illustrated by optical microscope studies of sample
sections. In this it is seen that because of the different shear rates of the
flow profile in a die, for example, near the surface of the moulded article,
typically an extrudate, the dispersed phase exists as extended ellipsoids,
whilst nearer the centre the particles are more spherical.[7] As will be clear
from the above, in so far as morphology can influence properties, the
wide range of possible morphologies implies in turn, a wide variation in
properties not only between different polymer types, but within a particular class of blend. Before discussing the individual properties of blends,
it is relevant to point out that although the particular property will have
a magnitude which frequently will lie between the values for the
individual polymers, cases are known in which the magnitude is less than

that of the individual polymers, and also where it is greater, this being an example of synergism.   This emphasises the role of one component in modifying the properties of the other, say the continuous phase component, and also in further interacting in a benevolent or antagonistic manner.

Considering first of all the thermal properties, as has already been mentioned, the presence of a single glass-transition temperature is an indication, no more, of a homogeneous system. More usually, certainly for amorphous polymers where the transitions are more prominent, two glass-transition temperatures are observed, with their locations varying from the values of the separate polymers to intermediate values as the compatibility or specific interaction of the two polymers is improved, until ultimately when complete compatibility is achieved, if this is possible, the two merge into a single peak (Fig. 6.5). The change in the

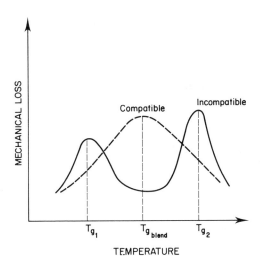

FIG. 6.5. Mechanical loss spectra of compatible and incompatible polymer blends.

behaviour will also be affected by the fineness of the dispersed phase, since this will lead to greater overall interaction in view of the greater total interfacial area. Turning to the case in which one of the polymers, at least, is crystalline, it appears that the melting point can be lowered by the other component, presumably through a chemical interaction effect, although this may be indirectly through a possible effect on the nature of the crystallisation process prior to measuring the melting point. In this connection it has been shown, for example, that in a polymer blend of polystyrene and crystallisable polyester, the polystyrene has a nucleating effect, the mechanism of which is not clear in that it may be a chemical effect or due to stress set up by differential thermal shrinkage behaviour.[8]

As far as optical properties are concerned, we have already seen that transparency depends both on miscibility and on particle size. If the latter is greater than the wavelength of light, scattering of radiation occurs and indeed can be used to study domain size, although this is not always easy even using laser light radiation. The principle of reducing particle size has been used with polyvinyl chloride/polybutadiene-styrene rubber blends to obtain transparent materials, but unfortunately this advantage was offset by a fall in impact strength. Attempts have been made to retrieve the situation by having larger particles, but using a third component.[9]

Turning to a consideration of the mechanical properties, the modulus will generally lie between the individual moduli of the two components. But the way in which it changes will depend not only on concentration but on morphology. For a homogeneous mixture the value will be a regular function not-too-far divorced from the proportionality relationship, but with heterogeneous systems the relationship can be quite complicated. In the case of a polyblend of say, rubber with a rigid plastic, some idea can be obtained from a Kerner-type analysis (see Chapter 3). Figure 6.6 shows the possible relationships.[10] Curves 1 and 2 show the upper and lower bounds for the cases where the two components are linked, with reference to a tensile stretching force, in a parallel and in a series juxtaposition respectively. The cases for dispersion of one component as spheres are shown at 3 and 4. For lower maximum packing fractions which also represent the situation for phase inversion the situation is as shown by curves 5 and 6. For IPN polyblends in which there is a continuous matrix for both components, the relationship will be as shown by the curve labelled IPN. Curves such as these can only provide a general guidance in view of the various and complicated phase and interactive relationships and the fact that inversion will depend upon relative viscosities at the moulding temperature. However, the results for polyethylene and butyl rubber confirm the expected trend. The way in which modulus changes with temperature will depend on the nature of the interaction and degree of compatibility between the phases. Figure 6.7 shows the typical behaviour for two amorphous polymers in a polymer blend for the cases of incompatibility and partial compatibility,

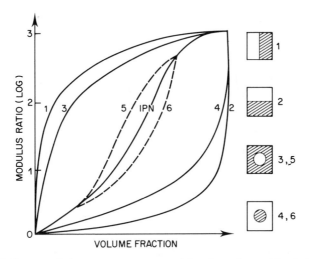

FIG. 6.6. Influence of phase geometry on modulus for polymer blends (see text). (Reproduced with permission of Marcel Dekker, Inc.[10])

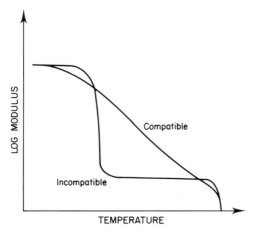

FIG. 6.7. Influence of temperature on the moduli of compatible and incompatible blends of amorphous polymers.

but again factors like the size of particles in the dispersed phase will play a part in influencing the actual shape of the curve. With reference to other tensile properties, Fig. 6.8 shows the effect of increasing the rubber content in a brittle polymer like polystyrene. Small amounts of rubber produce a yield point, this in practice being accompanied by crazing, but

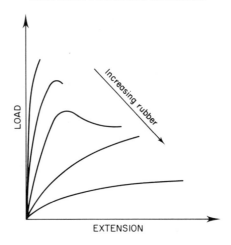

FIG. 6.8. Effect of the addition of a rubber on the load–extension behaviour of a brittle polymer.

as is seen there is an appreciable fall in tensile strength and extension of break. An exception to this trend in tensile strength has been reported for a polyurethane/polyacrylate IPN in which a synergistic increase in strength was observed.[11]

One of the main reasons for the success of polymer blends has been their impact resistant properties, where essentially a rubbery polymer is added to an otherwise brittle plastic. There is a reduction in modulus and strength and so a compromise must be sought, but generally the advantage of impact strength increase more than offsets the deterioration in these other properties, and additions of rubber up to about 30 per cent are acceptable. There has been considerable interest in the role of the second component in giving rise to improved impact behaviour and a number of possible reasons have been advanced.[12] Before discussing these it ought to be mentioned that if both components are rigid the impact strength of the blend is lower than that of either polymer. Where there is an improvement, as reported above, explanations centre on the ability of the rubber particle to absorb mechanical energy, the energy expended in craze formation, and the energy associated with debonding. It is now thought that the dominant mechanism is toughening by multiple crazing.[13] Whatever the mechanism in a particular case, there is no doubt of the efficacy of elastomers in producing this improvement in a wide range of polymers, including HIPS and polyvinyl chloride/ABS.

Cross-linked elastomers are better than non-cross-linked as they retain their identity on processing and allow fibrillation. As far as particle size is concerned, the impact strength falls off as size decreases below a critical particle size. This varies from system to system. For blends of styrene–butadiene block copolymers and polystyrene, the impact strength falls by a factor of × 7·5 as the particle size is reduced from 1 to 0·2 microns.[14] The impact strength of polyblends increases with temperature, particularly where a thermal transition takes place in one of the components.

Amongst other effects on polymer properties, mention should be made of the improvement in the friction characteristics brought about by the incorporation of low-friction polymeric additives. The influence is perhaps better than would be expected on a concentration basis since in practice the particles of the dispersed phase, typically polytetrafluorethylene, are constrained to lie nearer the surface rather than evenly throughout the matrix polymer.

The flow properties of polymer blends can be quite complicated, the overall flow masking the microrheological behaviour of the individual components. Depending on shear stress, for the same components, the apparent viscosity may increase or decrease with changing composition, possibly going through a maximum or minimum with respect to composition. Where an increase occurs, say in the addition of an elastomer to polystyrene in preparing a high-impact-resistant polystyrene, it is frequently necessary to decrease the molecular weight of the polystyrene to counteract the increased viscosity without appreciable loss in the impact behaviour of the final product. The influence on flow profile of the arrangement in space of the individual particles in a particle-filled polymer blend depends upon the size and deformability of the polymer filler, the smaller the size the less the effect on deformability. There is a tendency for deformable rubbery particles to move away from the wall of the moulding equipment, whilst for rigid spheres there is a trend away from both the wall and the centre of, say, the tube which can lead to layering of the dispersed phase. For the rubbery deformable filler, in perhaps an injection moulding process, there is a further tendency to elongation of the filler, this being greatest nearer the surface with an orientation of the major axis of the resultant ellipsoid in the direction of flow. Going towards the centre, the major axis turns towards the centre of the tube or channel, and in the centre the particles adopt an approximately spherical shape. Depending upon the rate of cooling of the composite, this being most rapid at the surface because of the low thermal conductivity affecting loss of heat from the centre, the individual deformations are retained. In extrusion it has been shown that relaxation of the elongated particles near the surface, before setting of the extrudate, can give rise to a matte (irregular) finish to the extrudate. From the technical point of view, polymeric additives are introduced to polymers during moulding to improve melt strength in blow moulding and thermoforming, and to reduce die-swell and melt fracture. In view of all the complexities to which reference has been made, it is difficult to calculate the viscosity of the fluid composite from a knowledge of the individual component viscosities. However, Nielsen[15] suggests as an approximation a form of the logarithmic Rule of Mixtures:

$$\log \eta_c = v_1 \log \eta_1 + v_2 \log \eta_2$$

Finally, let us look at some of the rather special applications of polymer blends. In addition to their perhaps major role in impact resistance, their ability to absorb energy has led to a possible use as 'silent paints', this time the energy being acoustic. They are finding an important application as polystyrene/polyphenylene oxide blends in which the high thermal resistance of the latter is balanced by the processibility of the former. Polymer laminates of two separate polymers have been used in film form for a number of years, having useful application in reducing permeation of the (protective) film to aggressive gases, liquids, and vapours. The advantage here may not only be in the corporate effect of two different materials each of which is active for different permeants, but also perhaps because of a non-linear dependence of diffusion coefficient on concentration which lowers the level of concentration of a diffusing substance at the surface of the second film. Another use of polyblends is in the addition of polyvinyl chloride to another polymer to improve flame resistance. Turning to polymer blends in the textile industry, in addition to the long-established process of physically blending staple fibres of more than one type, there is the possibility of internal blending produced by, say, conjoint spinning of two polymer melts. Examples of how this might be done are shown in Fig. 6.9. The composite, called a bicomponent yarn, will have some special properties derived from the different response of the two components to external agencies such as heating, solvent-swelling or drawing. For example, if the yarn is stretched, the recovery from deformation will be different for the different components, perhaps one taking permanent set. The result is a crimping effect as shown in Fig. 6.10, and has

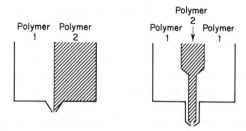

FIG. 6.9. Different methods for the spinning of bicomponent fibre.

(a)   (b)

FIG. 6.10. Crimping of bicomponent yarn. (a) As-spun; (b) cold-drawn.

therefore an application in simulated wool production. Other uses of blends are illustrated in the use of mixed polymers in tyres and in the recycling of the same, in the control of shrinkage in some thermosetting systems and in the reinforcement of soft polymers by, for example, the addition of phenolics, where an increase in the hardness and strength can be obtained.

BLOCK COPOLYMERS

In the same way that a polymer may be modified by means of a filler, the properties of the polymer may be alternatively modified by copolymerising it in the first place with another monomer, adjusting for example, the composition or sequence of placement of the second monomer unit to alter ultimate properties as desired. As a general way of modifying polymer properties this approach suffers from the same drawback as that associated with the introduction of a new homopolymer in that it requires the creation of a technology to itself as if it were a new polymer. Thus its special merit lies in its ability to provide some special property, perhaps permanence of composition, which is not readily available in a compounded polymer. On this basis, it would appear that the subject of copolymers has no particular relevance to the theme of this book, but in practice, the special area of block copolymers presents morphological and property implications which closely mirror those of polyblends, the main difference being that the phases which exist in block copolymer systems are linked from the outset by primary chemical bonds and so there is a mechanical continuity between different phases. Indeed, because of this close relationship, many reviews tend to make no distinction between the two.

The ways in which two monomer entities can combine are virtually limitless, but many of the variations lie within a number of basic possibilities. To exemplify, the following shows some of the ways that monomer units, M_1 and M_2, can combine:

$$\text{random} - \text{poly}(M_1 \text{co} M_2)$$
$$\text{alternating} - \text{poly}(M_1 \text{alt} M_2)$$
$$\text{block} - \text{poly}(M_1 \text{b} M_2)$$
$$\text{graft} - \text{poly}(M_1 \text{g} M_2)$$

The structural arrangement of these is shown in Table 6.2. In addition, there are star-shaped and comb-shaped copolymers, and there is the possibility of stereoisomeric sequences of the same or mixed monomer units. When one realises the further variations offered by differences in average molecular weight or weight distribution, some idea of the range of molecular structures and hence mechanical, physical, and chemical properties which are possible in these systems can be obtained.

For the present purposes we shall only consider two-monomer systems involving sequential runs of each, and thereby constituting basically block and graft copolymers. In the rare case that the two units are

TABLE 6.2

STRUCTURE OF COPOLYMERS

Code	Structure
Poly(M_1–co–M_2)	$-M_1-M_2-M_1-M_1-M_2-M_1-M_2-M_2-M_1-M_2-M_1-M_1-$
Poly(M_1–alt–M_2)	$-M_1-M_2-M_1-M_2-M_1-M_2-M_1-M_2-M_1-M_2-M_1-M_2-$
Poly(M_1–b–M_2)	$-M_1-M_1-M_1-M_1-M_2-M_2-M_2-M_2-M_2-M_1-M_1-M_1-$
Poly(M_1–g–M_2)	$-M_1-M_1-M_1-M_1-M_1-M_1-M_1-M_1-M_1-M_1-M_1-M_1-$

$$\begin{array}{ccc} M_2 & \qquad & M_2 \\ | & & | \\ M_2 & & M_2 \\ | & & | \\ M_2 & & M_2 \\ | & & | \\ M_2 & & M_2 \\ | & & \\ M_2 & & \end{array}$$

compatible there is the possibility of having a homogeneous polymer system which will have properties not too dissimilar to those of random or alternating copolymers. But in the more common case where the sequences, being representative of different chemical types (since otherwise there would not be a convincing reason for making the polymer in the first place), give rise to some mutual rejection, a heterogeneous system will be formed. Just as there is a range of different types of copolymers, there is also within the group of block copolymers a range of possibilities, each one giving rise to different properties. One example is a block copolymer of the following sequence:

$$(M_1 \ldots M_1)(M_2 \ldots M_2)(M_1 \ldots M_1)$$

which, representing a polymer of the M_1 species as A and a polymer of the M_2 type as B, can be alternatively characterised by the notation:

ABA block copolymer or ABA poly($M_1 b M_2$)

Other examples of related block copolymers, expressed in the same code, are AB, ABAB, $(AB)_n$, and BAB copolymers. From the point of view of microstructural arrangement, the properties of all these will be different, but can be further emphasised or moderated by preparative and fabrication variables. It is not intended to discuss here the various ways in which block or graft copolymers can be made, except to say that these depend to some extent on the specific nature of the reactants. The more important ways are step-growth polymerisation, mechanical breakdown with molecular rearrangement of mixed homopolymers or of one homopolymer in the presence of a reactive monomer, anionic polymerisation, and Ziegler–Natta polymerisation. In addition there are methods based upon cationic polymerisation, controlled silane reactions, biochemical enzymic synthesis, etc. Of all these it was probably the discovery of the preparation of 'living polymers' by anionic polymerisation in which growth of a separate polymeric species at the end of an active free radical is arranged to take place, which has had the most dramatic influence on the preparation of block copolymers. This permits chain extension where an alternative approach by use of terminally active chains has not proved possible because of incompatibility and end-group protection, presumably through chain coiling when two polymers are physically mixed. By these various techniques a number of important commercial block copolymers have been produced. These include ABA, AB, and BAB polymers of styrene and butadiene or isoprene, segmental polyurethanes, ethylene–vinyl acetate (EVA) copolymers, polystyrene–polysiloxane polymers, and block polyesters. Some of these, like the ABA poly(styrene-b-butadiene) and some of the polyurethanes find application as thermoplastic elastomers, i.e. materials which have thermoplastic properties and yet behave as elastomers, whilst others are used as surface-active agents for oil-based polymer dispersions, and as coupling agents for polymer blends.

As already mentioned, the properties of the different copolymers will depend not only on the chemical and stereochemical nature of the monomer units but also on copolymer sequence, distribution, length, and number of blocks in a polymer chain. These, with copolymer composition and solid-state morphology will all contribute to ultimate properties, some being interrelated. As might be imagined, a complete theoretical analysis of these and other variables is both difficult and complex, and assessment of the predictions by structurally analysing a solid polymer is not at all easy. However, attempts to do both have been carried out and represent an area of commendable effort. From the theory it appears that the tendency to phase separation decreases as the number of blocks in a given chain length increases and for a constant composition and a given number of blocks in a chain, the greater the molecular weight of a given block the more likely is the system to phase-separate. The actual organisation of the two phases depends upon the relative concentration of the two molecular species, but other factors such as the chemical nature of the species and their relative viscosities at the moulding temperature will also affect the separation. The general trend is as shown in Fig. 6.11 where A would typically be a soft elastomeric polymer type and B a hard plastic related phase or domain. The actual structure of the domains is not always identifiable but some

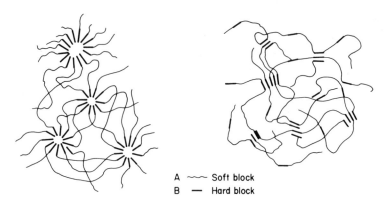

A ~~~ Soft block
B — Hard block

FIG. 6.11. Phase organisation in block copolymers.

general guidance can be provided. It appears that for an ABA block copolymer of styrene–butadiene–styrene, say in a typical composition found commercially, the polystyrene blocks associate because of compatibility and thus act as physical tie-points in an otherwise flexible matrix. The size of the hard block domains can be calculated, and for a copolymer of polystyrene blocks of molecular weight of about $10–15 \times 10^3$ and polybutadiene blocks of about $5–7 \times 10^4$, the domains are beginning to approach the wavelength of light and so are able to scatter light. The sizes of the domains are believed to be less than for polyblends of the same molecular weight. The identification of hard blocks in polyurethanes, being isocyanate-chain extender sections, and soft blocks being typically polyester or polyether sequences, with the corresponding hard and soft domains is not believed to be as distinct as for the polystyrene–polybutadiene block copolymers and the size of the domains is believed to be smaller. A further sophistication can arise with polyurethanes in that a separate morphology can develop where either one or both of the blocks is capable of crystallisation.

As has already been stated, the nature of the morphological organisation depends upon fabrication conditions especially with respect to temperature which will not only affect the relative viscosity of the two components during moulding but also the extent of specific interaction between the two. If the copolymer is prepared by casting from solution, depending on the preferred compatibility of a solvent for one or other of the blocks, phase inversion phenomena, as for polyblends, is possible. Some influences of the nature of the casting solvent on the properties of a block copolymer together with corresponding data for melt-fabricated polymer are reported in Table 6.3.[16] Any stretching of the copolymer which leads to permanent set or partial domain breakdown, possibly recoverable on annealing, will also affect morphology and hence

TABLE 6.3

INFLUENCE OF PREPARATION CONDITIONS ON THE MECHANICAL PROPERTIES OF A BLOCK COPOLYMER (POLYSULPHONE–POLYDIMETHYL SILOXANE)

Preparation method	Modulus (MN/m^2)	Tensile strength (MN/m^2)	Elongation (per cent)
Tetrahydrofuran-cast	245	22·8	410
Benzene-cast	154	20·8	517
Ethyl acetate-cast	61	13·0	250
Moulded (320°C)	57	11·9	70

mechanical and other properties. Stress softening is an example of such an effect. Some typical stress–strain curves for common block copolymers are shown in Fig. 6.12. Curves of this kind can also reflect the different influence of different sequence arrangements of blocks in the chain, an AB or ABA polystyrene–butadiene polymer having for the same composition ratio of the two species, a softer constitution because of lack of structural continuity through hard block domains, the terminal sequences now being of the less strongly interacting polybutadiene

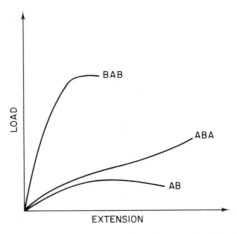

FIG. 6.12. Load–extension behaviour of different A/B block copolymers.

blocks. In the light of all these variations, although Kerner-type equations have been used to describe the mechanical properties of discrete dispersed materials, such a use must be treated with care. Other factors which bear on mechanical properties are the molecular weights of the blocks, and any branching or cross-linking. In general, except at low molecular weights which can affect compatibility or, with crystallisable blocks, the extent of crystallisation,[17] the length of the block is not of importance. Branching within a given molecular weight will decrease tensile strength, whilst cross-linking, except perhaps where it might interfere like branching with any tendency to crystallise, will generally increase stiffness although the presence of a filler may affect the level of cross-linking.[18]

The thermal properties of block copolymers differ from those of other copolymers as is seen in Fig. 6.13(a). The melting points are not

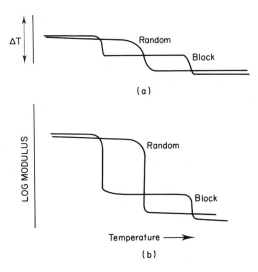

(a)

(b)

FIG. 6.13. Thermal and mechanical behaviour of block and random copolymers.
(a) Glass-transition changes in DTA; (b) modulus changes.

substantially different from the values of the separate homopolymers, bearing in mind any molecular weight differences for those copolymers which have one or two crystallisable units. A similar behaviour is observed for glass-transition temperatures, the individual values for the two homopolymers being retained when the two blocks are incompatible. Both transitions can be detected by differential thermal analysis and by dynamic mechanical analysis, the main difference in the latter case from that of polyblends being that since molecular continuity exists across the phase interface, one does not get as is sometimes found for polyblends, the disappearance of one of the peaks. The influence of the separate transitions, say on mechanical properties of a copolymer with largely amorphous components, is shown in Fig. 6.13(b). If there is some degree of compatibility, then the fall in modulus, for example, is more gradual with respect to temperature than otherwise would be the case. The influence of temperature also reflects a difference in the behaviour of thermoplastic elastomers of the block copolymer type compared with conventional elastomers of the thermosetting type. In the latter, as the temperature rises, the increased mobility of the flexible chains produces a tension on the tie points which for the extended polymer produces a stiffening of the test sample, i.e. as distinct from other materials not undergoing any other structural change, the modulus increases with temperature. Only when the temperature rises to such a level that chain breakdown occurs does the modulus start to fall again. For thermoplastic elastomers, as the temperature rises, the increased mobility leads to some domain breakdown. Thus the effective number of tie points decreases and so the overall mobility now leads to a decrease in the modulus. By adjusting the composition of a thermoplastic polyurethane elastomer, in an attempt to balance the stiffening effect between tie points and the tendency to breakdown of tie points, it has been possible to prepare a polymer whose modulus–temperature relationship remains fairly constant over a range of temperatures.[19]

It was mentioned above that the molecular weight of the blocks for high molecular weight sequences did not have a dramatic effect on the mechanical properties, but for the flow properties this is not the situation, viscosity increasing with molecular weight. In general, the viscosity of block and graft copolymers is higher than expected on the basis of total molecular weight and composition where the blocks are incompatible. The linkages between the blocks mean that molecular integrity is maintained upon deformation in the melt and this can have an influence on the morphology leading to lamellar domains. Interaction between blocks of the same type on different chains, at the secondary bond level, is considered to lead to some weak but positive structure formation in melts which can persist to low rates of shear.

The theme so far in this section has been to emphasise the resemblance of block copolymers to polymer blends and in turn to establish them as being essentially a type of polymer composite. In addition, however, the polymers themselves may be further compounded with fillers in order to modify properties, or to lower the cost. Indeed, large quantities of inexpensive fillers are used, sometimes in cooperation with plasticisers to offset the stiffening effect of fillers, both for modification of applicational properties and flow. Fillers can be useful in providing an increase in strength at higher temperatures. Examples of fillers used in this way are silica, carbon black, and clays which improve tear strength, flex life, and stiffness. Some fillers like titanium dioxide and carbon black help to shield against UV breakdown as they do for conventional polymers, and the block copolymers may be blended, not only as compatibilising agents but in their own right, with other polymers to form polyblends. Examples of this are in the addition of polystyrene to styrene–butadiene copolymers to improve tear strength, flex life, and hardness, and the blending of polyethylene and ABS resins.

REFERENCES

1. FLORY, P. J., *Principles of Polymer Chemistry*, Cornell U.P., Ithaca (1953).
2. BUCKNALL, C. B., *Toughened Plastics*, Applied Science Publishers, London (1977).
3. KRAUSE, S., in Paul, D. R. and Newman, S., (Eds), *Polymer Blends*, vol. 1, Chapter 2, Academic Press, New York (1978).
4. ICHIHARA, S., KOMATSU, A., and HATA, T., *Polymer J*, **2**, 640 (1971).
5. ZAKIKHANI, M., PhD Thesis, University of Bradford, in prep.
6. THOMAS, D. A., and SPERLING, L. H., see ref. 3, vol. 2, Chapter 11.
7. GOLDSMITH, H. L., and MASON, S. G., in Eirich, E. (Ed) *Rheology*, vol. 4, Academic Press, New York (1967).
8. SHELDON, R. P. and WRIGHTSON, I., *Brit. Polymer J.*, **10**, 215 (1978).
9. STEIN, R. S., see ref. 3, vol. 1, Chapter 9.
10. NIELSEN, L. E., *Mechanical Properties of Polymers and Composites*, vol. 2, Marcel Dekker, New York (1974).
11. FRISCH, K. C., KLEMPNER, D., FRISCH, H. L. and GHIRADELLA, H., in Sperling, L. H. (ed) *Recent Advances in Polymer Blends, Grafts and Blocks*, Plenum Press, New York (1974).
12. Ref. 2. p. 188.
13. BUCKNALL, C. B. and SMITH, R. R., *Polymer*, **6**, 457 (1965).
14. DURST, R. R., GRIFFITH, R. M., URBANIC, A. J. and ESSEN, W. J. VAN, *Am. Chem. Soc. Adv. in Chem. Ser.*, **154**, 239 (1976).
15. LEE, B–L. and WHITE, J. L., *Trans. Soc. Rheol.*, **19**, 481 (1975).
16. MARK, H., GAYLORD, N. G. and BIKALES, N. M. (Eds) *Encyclopedia of Polymer Sc. and Tech.* suppl. 2, p. 147, Interscience, New York (1977).
17. SUCHIVA, K., PhD Thesis, University of Bradford (1978).
18. MARASHI, M. M. M., PhD Thesis, University of Bradford, in prep.
19. MARASHI, M. M. M. and SHELDON, R. P., *Brit. Polymer J.*, **11**, 89 (1979).

Chapter 7

COMPOSITE SYSTEMS

INTRODUCTION

Having seen something of the fundamental nature of composite polymeric systems, their formulation and methods of production, general mechanical properties, and their more important physical and chemical properties, we are now in a position to review the range and application of the major representatives. To do this one could choose either the fillers or the polymers as a base for assessment. In what follows the latter approach has been adopted. Reinforcement and the use of fillers is of particular significance with about twenty different major plastics, in addition to natural rubber and other (synthetic) elastomers. Of the plastics, polyvinyl chloride, polyesters, phenolic resins, polypropylene, epoxy resins, and nylon account for about 95 per cent of the total.[1] Although polyvinyl chloride is probably the most important in terms of tonnage, polyester resins are catching up, whilst polypropylene has the highest growth rate. These and other examples will be discussed below. But before doing this, some mention should be made of other topics to be outlined in this chapter. These include materials, which although technically polymer composites, have a character outside the main theme so far reported. The first is that of reinforced expanded polymers in which the primary filler is gaseous, but to which further reinforcing solid filler is added. The second concerns the subject of polymer cements in which the primary phase is inorganic and so can be regarded as a phase-inverted polymer composite with properties largely dictated by a cement matrix rather than an organic polymer. Alternatively, they could be regarded as inorganic polymer composites. Another reason for including

178

them here is to emphasise the fact that polymer composites are only a part of the bigger subject of composite materials in general. A growing area in reinforced polymers involves the use of hybrid fibre systems, and so a section is devoted to this topic. Finally, an attempt is made to delineate trends in composite polymeric systems so that some future projection of their importance can be made.

THERMOSETTING POLYMER COMPOSITES

The characteristic property of this class of polymer system is, as indicated in Chapter 2, that they are chemically cross-linked prior to application. Thus compounding must be carried out either before or during polymerisation since the composite cannot be reshaped without degradation. On the other hand, because of the rigidifying effect of both filler and cross-linking, not only is it possible to achieve high stiffness, but the material, once formed, is fairly insensitive to temperature, except at the point where the polymer begins to break down. In terms of overall production, this class of polymers is less important than that of the thermoplastics, although at the present time the latter group of polymers is marketed mainly without reference to the incorporation of fillers. Nevertheless, as already mentioned, this area is of very great importance in the context of reinforced thermoplastic engineering materials, competing, as we have seen, with traditional materials of construction, which also incidentally are experiencing an upgrading in properties for basically the same reason as are polymers. One of the significant deficiencies of the thermosetting resins from a commercial point of view is that production tends to be more labour-intensive. Despite this, great strides have been made both in modifying the systems so that they can be accommodated by production techniques normally associated with thermoplastic polymers, and in developing new methods of fabrication which allow thermosetting polymers to be moulded in ways competitive with the above. A case in point is reaction injection moulding which was described in Chapter 5, this being only one of a number of advances which have been made in recent years towards a more automated approach to thermoset processing. The following outlines some of the more important polymeric composites and in so doing follows the general sequence of polymers described in Chapter 2.

The first and originally the most important family of composites was that based upon phenolic resins. It should perhaps be pointed out at this stage that, although certain fillers are preferred for specific purposes, there is considerable latitude in the nature of the unfilled resin, so that a double advantage is frequently possible in some direction such as improved electrical resistivity or thermal stability through the choice not only of the best filler, but also the best resin in the first place. The same principle will apply to many of the thermosetting polymers and some of the thermoplastics, although in the latter case, a new polymer, even though related to another polymer, may well require a new technology and as such may well constitute essentially a new class in itself. The general-purpose phenolic moulding compounds are usually based on purified and finely divided woodflour, derived from softwood trees such as pine, spruce, and poplar, the coarser untreated wood chips being used as the filler ingredient of chipboard. Woodflour provides a cheap filler which improves impact resistance and like many other cellulosic materials seems to have a natural affinity for phenolic resins. The filler also reduces both shrinkage and exotherm. Better mechanical properties, including impact resistance, are obtained through the use of fibrous fillers. These include cotton, sisal in the form of flock for example, chopped fabric, cord or string, and paper pulp, giving impact strength increases of up to five times that of the unfilled resin, and nylon fibre and fabric, rayon, and glass, which give increases of about twenty times. Asbestos fibre also improves impact strength but has the further advantages of heat and electrical resistance. Mica, however, is the filler most usually used in electrical-grade phenolics. Mention should be made of the use of asbestos and also silica and quartz fibres in ablative applications. Metal fillers find use in phenolics for radio and electrical components, with lead having a special role in some medical applications where X-rays are concerned. Although the properties of filled phenolic resins will of course vary with loading, in practice compositions in which the weight ratios for organic fillers are about 1:1 and inorganic fillers about 1·5:1, are typically used. General-purpose phenolics find wide application in the electrical industry, and as cases, knobs, bottle tops, handles, etc. Impact-resistant grades find use in gears, tool handles, washing-machine parts, and industrial parts generally. Heat and electrical grades are used in electronic and electrical engineering and in the automobile industry. The other major area of filled phenolics is that of the laminates which are based on impregnated paper, cotton, glass, nylon, asbestos, etc., and are used in the fabrication of sheets, rods, bars, and tubes. They find application in high-voltage insulation, gear wheels, and bearings as well as for chemical resistance purposes. Sometimes, instead of being used in laminating, the impregnated sheets may be dried and then granulated for moulding high-impact-resistant articles.

The urea–formaldehyde and melamine–formaldehyde aminoplast resins, although more expernsive, except for a few urea–formaldehyde systems, and having possibly inferior impact behaviour, form optically lighter and clearer materials. For this reason in particular, and because of better electrical arc resistance, due to a lower tendency to carbonise, they find special application where these two attributes are required. The range of commonly used particulate fillers is more restricted than for the phenolics, with bleached and purified wood pulp being the preferred filler. Fibrous fillers such as glass, asbestos, and cotton fabric also find wide application. Melamine–formaldehyde composites have an advantage over the urea–formaldehyde systems in that they have lower water absorption, somewhat better electrical properties, better heat resistance, stain resistance, and hardness, the last three properties leading to extensive use in tableware. The further advantage of improved colour compared with the phenolics, permits the use of decorative printed surfaces, the principle being used in decorative laminates. In these latter applications, internal layers may be of the cheaper, but darker, phenolic laminates since these are not visible. Despite the advantages of melamine-based systems they suffer strong competition from epoxy resins and polyesters. Aminoplasts are used in bottle tops, buttons, and knobs, and particularly in electrical products such as switches and other objects.

One of the biggest and most important groups of polymer composites, both thermosetting and thermoplastic, is that based on the unsaturated polyesters, including alkyd resins and the diallyl phthalate and iso-phthalate related polyesters. There are many reasons for the present popularity including good mechanical, physical, and chemical properties, ease of production, versatility in formulation, and adaptability of production process to a given application. Even without the use of a selected filler or reinforcing agent for some particular application, there is great scope in the choice of chemical component for the best electrical characteristics, thermal stability, refractive index, and flexibility, as was outlined in Chapter 2. Then there is the further potential, particularly with respect to some optimum mechanical property, not only from the available range of reinforcing fibres, glass being the most important at the present time, but also from the numerous different geometrical arrangements in space of the fibres, dictated to some extent by a given production or fabrication process. The filler can either be particulate, or, as mentioned above, fibrous. Discussing first of all, the particulate class of fillers, these are incorporated because of their effect either on processing or on final properties. For the first case, they are used to increase viscosity, produce thixotropy, reduce exotherm, affect rate of cure, etc., whilst for the latter, they influence thermal expansion, increase rigidity, impact strength and heat distortion temperature, affect moisture uptake, chemical resistance and electrical behaviour, and by reducing shrinkage on cure, have an influence on those mechanical properties which are sensitive to this shrinkage. In both areas of application, whether processing or property modification, cost and resin compatibility, particle size, and concentration are all important. Such a wide range of considerations suggests a correspondingly wide range of different fillers, and indeed, this is the case with some forty to fifty particulate fillers commonly being used in polyesters[2] (see e.g. Table 7.1). As already noted, the main fibrous filler is glass, but carbon and carbon–glass hybrids are of growing importance. The glass which is used is available in many forms including chopped strand mat, continuous strand mat, roving, woven roving, cloth of different weaves and density, and continuous filament, all of which may be used separately or in combination. The various forms have a bearing on the fabrication processes, which again have already been described, in Chapter 5, and include hand and spray lay-up techniques, vacuum bag moulding, pressure bag moulding, injection moulding, centrifugal casting, filament winding, and pultrusion. It is quite clear that for the production of some particular article, in many cases more than one of these methods could be employed, so that the decision as to which method should be chosen might rest on such considerations as cost, convenience, and any special property required. In making this decision factors like length of run, cost of materials and process, product size and shape, any need for further trimming or finish, etc., would have to be taken into account, but generally for short runs, hand lay-up would be seriously considered. For intermediate volume productions cold press and resin injection would claim attention and for large overall volume fabrication, hot press moulding would be of importance. For high directional strength requirements, filament winding and pultrusion would have to be examined. To illustrate the typical result of these deliberations, reference should be made to Table 7.2 which indicates the products associated with the different processes.[2] To conclude, the

TABLE 7.1
EXAMPLES OF PARTICULATE FILLERS USED IN POLYESTER RESINS

Filler	Oil absorption (g/100 g)	Main uses
Aluminium trihydrate	20–40	Flame retardancy
Alumina	10–50	Abrasion resistance, dimensional stability, electrical and thermal properties
Antimony oxide	10–15	Flame retardancy
Barium sulphate	5–15	Pigment
Bentonite	30–55	Thixotropic agent
Calcium carbonate (ground)	5–15	General purpose, pigmentation, dimensional stability
Calcium carbonate (precipitated)	15–60	General purpose, pigmentation, dimensional stability
China clay	25–45	Pigment, thixotropic agent
Glass spheres (solid)	—	General purpose, chemical and moisture resistance
Glass spheres (hollow)	—	Light-weight filler
Graphite	—	Thermal and electrical conductivity, pigment
Magnesium oxide	55–70	Thixotropic agent
Mica	50–70	Chemical and moisture resistance, electrical and thermal properties
Phenolic microballoons	—	Light-weight filler
Silica	15–40	General purpose, abrasion resistance, dimensional stability, electrical and thermal properties
Talc	20–45	General purpose
Zinc oxide	10–15	Pigment
Zirconium silicate	—	Abrasion resistance, chemical and moisture resistance, electrical and thermal properties, dimensional stability

TABLE 7.2

TYPICAL PRODUCTS OF POLYESTER/EPOXY MOULDING PROCESSES

Process	Products
Hand lay-up	Boats, building panels, general
Spray lay-up	Boats, building panels, general
Vacuum/pressure bag	Aircraft sections, panels, general
Foam reservoir	Automotive, furniture, etc.
Resin injection/ resin transfer	Boats, etc.
Vacuum impregnation	Radomes, aircraft nose cones, etc.
Cold compression	Automotive, industrial, electrical
Hot compression	Automotive, industrial, electrical
Transfer	Small to medium sized components
Injection	Small to medium sized components
Filament winding	Tanks, pipes, tubes
Centrifugal	Pipes, tubes
Continuous sheet	Roofing lights, etc.
Pultrusion	Rods, tubes, etc.

applicational areas of reinforced polyesters in order of output importance should be noted. These are transportation, marine applications, construction, corrosion resistance, electrical goods, consumer goods, aircraft and aerospace, and, finally, a range of miscellaneous applications. Under these headings would come such diverse examples as boat and even warship hulls, roof lights and building cladding, light-weight car bodies (currently of particular importance in the context of energy saving), bathroom fittings, buttons and other small item mouldings, and various products which extend the use of polyesters to an art form.

The next most important thermosetting polymer and one which challenges polyesters in a number of applications, is epoxy resin, which has advantages of durability, low shrinkage, good adhesion to other materials including metals, and good heat and chemical resistance. As with the polyesters a wide range of reagents afford a correspondingly wide influence on properties, even before the incorporation of filler. Several hundred fillers have been recommended for use in epoxy resins. In the case of particulate fillers, little use is made of organic fillers although phenolic microballoons have been used for cores. On the other hand much use is made of inorganic materials for improving rigidity, stability to heat, chemicals and radiation, etc., and for providing good mechanical and electrical properties. They include calcium carbonate as marble flour and chalk, sand and silica flour, mica, slate, vermiculite, zircon, and aluminium. Some care must be taken in matching filler to a particular epoxy composition, in addition of course to any surface pretreatment. For example, mica, which is widely used, can give rise to undesirable reaction with some anhydride curing agents in the same way that asbestos can unfavourably interact with boron trifluoride catalyst complexes. Certain fillers are chosen for specific purposes. Typical of these might be aluminium and copper metals for improving heat conduction, alumina for surface hardness, silicon carbide for abrasion resistance, molybdenum sulphide, calcium carbonate, and aluminium, etc., for machinability as compared with silica- or alumina-filled resins which are more difficult to machine. One of the attributes of epoxy resins is their low coefficient of expansion, which can be further reduced by incorporation of mineral fillers, one of the most effective being zirconium silicate which itself has almost zero thermal expansion coefficient. Of the fibrous fillers, glass fibre is the most important after it has been suitably treated to make it compatible with epoxy resins. As for the polyesters, a wide range of fabrication techniques is available, the standard technique being lamination using cloth, scrim, mat, etc., but filament winding, pultrusion, and vacuum impregnation techniques are also well used. Carbon fibre is finding increasing use either separately or in hybrid combinations with glass fibre. Some of the newer fibres such as boron and aramid fibres are also used with epoxies and appear to be preferred for this polymer rather than with polyesters. Amongst the applications of filled and reinforced systems are glass and carbon fibre structures in electrical, nuclear, mechanical, and aerospace engineering. In the last respect it is perhaps fitting that epoxy resins continue in the same tradition established by their use as adhesives in aircraft, although today they find more sophisticated application as the basis of propellor blades using carbon fibre reinforcement and as the material for the cargo door of the Space Shuttle, offering a weight saving of 25 per cent compared with an alternative light alloy. The characteristic nose cone of Concorde is a reinforced glass fibre epoxy resin being transparent to radar and able to withstand supersonic stresses and temperature changes of between -50 and $190°C$, and carbon fibre reinforced epoxy resin fuselages are being used for the construction of aircraft which actually weigh less than their fuel load. Other applications of reinforced epoxy resins include tooling, chemical resistant flooring, and electrical goods including printed circuits.

Composites based on silicone polymers tend to be expensive compared with those previously described, but on the other hand they have some special areas of application of some considerable importance. They are available as moulding compounds, laminates, and elastomers, particularly room-temperature-vulcanising rubbers (RTVs). In the first category, mineral and short length fibres are used for reinforcement and filling of electrical component encapsulating resins which find high-temperature applications in aircraft and missile parts. Glass cloth is the principal laminating filler, again in electrical applications of silicone resins where they are generally superior to phenolic and melamine formaldehyde composites, although interlaminar shear properties are frequently inferior to these and to epoxy resin composites. Mica and asbestos find use in the same area, but to a lesser extent. Silicone elastomers often incorporate finely divided silica as reinforcing filler having a vulcanising action as well. As far as general-purpose polymers are concerned, they find use because of good thermal and electrical properties, principally in aerospace applications as seals, gaskets, ducting, etc. They are also employed in marine cabling. A final important application of filled silicones is in connection with aluminium-filled surface coatings for metal chimneys and furnace doors, which leave a protective film of metal after the silicone has broken down under the high temperatures.

Although fillers, especially pigments, are added to polyurethane polymers, the properties of these, like many of the above polymers but to a greater practical extent, can be considerably modified internally by correct choice of reactants, relative concentration, and preparative conditions. They find use in composites as additives to other polymers but the main area of current interest is that of RRIM composite polyurethanes, involving the addition of hammer-milled and chopped-strand glass fibre to systems used to produce microcellular polyurethane. The main outlet for these products is in automobile parts, but with more recent extension to furniture, domestic appliances, and for general moulding. The advantage of the filler is in increasing flexural modulus and providing better dimensional stability over a range of temperature without too severe loss in impact strength and elongation at break.

Although there are other thermosetting polymers of interest, this section concludes with reference to only one class, that of cross-linked elastomers. Some mention has already been made of filled silicone rubber and polyurethane polymers, some of which are both lightly cross-linked and elastomeric, but it should be realised that almost all elastomers contain a filler amongst the list of compounding ingredients. Of these carbon black is the most important, not only in natural rubber but also in synthetic elastomers like styrene–butadiene rubber (SBR) which being non-crystallising would otherwise lack the high extension reinforcement, characteristic of the crystallising elastomers. For natural rubber, carbon particulate filler is introduced at a level of 25–400 phr for the various applications which it has, including the most important in terms of volume production, that is automobile tyres. The three main reasons for use of carbon in natural rubber are its reinforcing action, pigmentation, and electrical conductivity, the latter of special importance in eliminating the build-up of static electricity. Silica is another important filler, but requires pretreatment with propylene oxide or a silane to improve compatibility. Other fillers used to extend the properties and therefore not necessarily added for reinforcement, include zinc oxide, having the triple role of filler, pigment, and curing accelerator, and which indeed, was used before carbon in tyres, talc, barium sulphate, kieselguhr, calcium carbonate as whiting, clays which increase hardness and tear strength and in the case of red clay also tensile strength, hydrated alumina which assists processing and improves mechanical properties, and a range of silicates being less effective than hydrated silica, but better than clays and whiting. Amongst the few organic fillers which are used, mention should be made of phenoplasts, coumarine–indene resins, lignin, etc. Cellulosic fibrous reinforcement derived from wood pulp produces an increase in modulus at the expense of an initial decrease in tensile strength, which, however, increases again to a maximum at about 25 per cent concentration, but for the best properties, surface treatment of the filler is necessary[3] (see Fig. 7.1). Powder and latex blending are two techniques which lend themselves to additional ways of introducing fillers, and latex blending is particularly important for the compounding of rubber with paper and woven textiles. The other area of elastomer-based composites which is of importance is that which usually comes under the heading of toughened plastics. Something has already been said about these in the discussion of polymer blends in which, typically, 5–10 per cent rubber is dispersed in a more rigid plastics matrix. Natural rubber was used in this way as an impact-improving additive to polystyrene many years ago, although SBR is more compatible and is more often used nowadays. ABS is another essentially reinforced (i.e. for impact behaviour) polystyrene plastic with impact strength and toughness increasing by about a factor of 7 for normal levels of addition. Polyvinyl chloride is reinforced with ABS and other speciality elastomers at about the 5 per cent level, and polypropylene also may receive natural

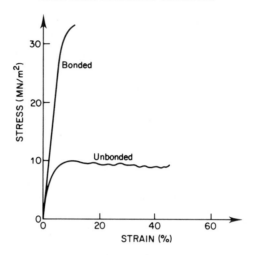

FIG. 7.1. Influence of bonding agent on the tensile properties of cellulose-
reinforced natural rubber.

or synthetic rubber incorporation. For example, it is typically blended
with ethylene–propylene rubber to improve impact strength; these blends
which have slightly lower shrinkage than polypropylene itself are finding
increasing use in the automobile industry. In addition to polypropylene,
epoxy resins and to a slight extent, even polyethylene are also compounded with elastomers. In order to retain the transparency of polymethyl methacrylate, at least over a small temperature range, and yet to
increase impact behaviour, use can be made of speciality elastomers
based on methyl methacrylate–acrylonitrile–butadiene–styrene elastomer.

THERMOPLASTIC POLYMER COMPOSITES

Thermoplastic composites are usually divided into two classes. The first
is the long-established group of mainly particulate-filled polymers in
which the filler is present primarily as a diluent, to reduce cost without
too serious an effect on useful properties. Occasionally, an added stiffness
or wear resistance is the main reason for use of a certain filler. The
second class of composites is that of the reinforced engineering thermoplastics. In practice, real systems are frequently compounded of a mixture
of the two classes so that some optimisation of properties is achieved.

For the purposes of the present section, each class will be discussed in turn.

Particulate-filled Thermoplastics

Although many thermoplastic polymers such as polymethyl methacrylate, polystyrene, and polycarbonate find particular application because of their transparency, probably all plastics at some time are marketed in filled form, although the filler may be present as a pigment rather than as a means of improving some property or for reducing cost.

Since the number of fillers which can be used is very great, as is the range of individual polymers and their commercial grades, only a broad survey of particulate-filled polymers can be given here, and so this account will be restricted to the so-called commodity, or large-volume production, polymers with some mention of semi-commodity polymers since some of these will possibly reach the higher status in the coming years. It should be realised, however, that a dramatic effect of some filler on a new or speculative polymer, especially if the filler can be introduced usefully at a high concentration which might offset a cost disadvantage, might allow it to compete with one of the better-known thermoplastic polymers. If the influence on properties is very special as for aerospace or medical application, then a hitherto unknown polymer may become of considerable importance. It is hoped that this survey will at least indicate the potential of the latter possibility, by providing some understanding of the effects of fillers on the better-known polymers.

Beginning with polyethylene, carbon black is probably the most commonly used filler both because of its pigment role and also because of its ability to act as a UV stabiliser, but clays, titanium dioxide, zinc oxide, iron oxides, and other pigment fillers are also used. In addition finely divided silica increases both heat distortion temperature and modulus, with pyrogenic silica having the greatest effect on the former property, raising it to 77°C compared with 54°C for carbon black at the same loading of 50 per cent. Talc at a level of 40 per cent increases flexural modulus of medium molecular weight polyethylene by a factor of 3, but usually, levels for optimum influence on heat distortion temperature are less than this, say about 30 per cent. Calcium carbonate has an optimum influence on toughness and softening point at about the same concentration. Barium sulphate and clay do not produce significant improvements in mechanical properties but are useful in reducing mould shrinkage. Maleic anhydride, polymerised in the presence of clay-filled polyethylene, can produce an improvement in mechanical properties.

Another filler worthy of mention is starch used at a level of about 30 per cent, the composite being used as described earlier as material for biodegradable 'throw-away' plastic bags, etc., competing with related composites which contain UV sensitive chemicals to promote light degradation. The copolymer of ethylene, ethylene–vinyl acetate polymer (EVA) is frequently sold as a composite containing precipitated silica, or alternatively with carbon, in the latter case for electrical purposes.

Talc is used as a filler for polypropylene, especially for automobile parts where the improved surface finish derived from the presence of the filler, ease of processing, and low mould shrinkage are valuable in the production of heater housings, ducts, and pump parts. Compared with calcium carbonate, there is an increase in modulus without serious deterioration in tensile properties, although this system also finds application in automobile parts. The use of filler can produce a decrease in high-temperature stability which may be moderated through the further use of stabilisers.

The thermoplastic polymer which accounts for the use of a great deal of fillers incorporated into polymeric systems is polyvinyl chloride, the filler most used being calcium carbonate. The largest proportion is mixed with flexible (plasticised) polymer in the concentration range 20–60 per cent phr. Calcium carbonate improves stiffness, tensile strength, and tear strength, and so is often used in flooring grades of the polymer. Asbestos has been the next most important filler for polyvinyl chloride, followed by talc, clays, and silica. Finely divided silica and silicates are compounded with PVC for flexible sheeting to improve handle and drape, and at the same time processing and the tack of plasticised formulations is improved. Plasticised polymer containing silica is used also in shoe soling. Pyrogenic silica at a level of about 10 per cent can increase the heat distortion temperature by about 30°C and will improve electrical properties. For use in PVC, silica is generally pretreated with a silane coupling agent. Other fillers include antimony oxide, which helps to improve the flame resistance, carbon black for improvement of weathering, mechanical properties, and electrical conductivity, this being important for PVC belting for use in mining, to remove static electricity, zinc oxide as a heat and light stabiliser, and alumina, barium sulphate, and gypsum used as general-purpose fillers.

Polystyrene is not normally regarded as a compounded polymer since it is the inherent clarity of the polymer which leads to a number of its applications, but pigments are used, and talc will increase stiffness, although the polymer is usually regarded as a rigid material in its own right. Woodflour is another filler which is occasionally used. However, the most common compounding situation for polystyrene, as has been mentioned a few times, is in connection with its use in polymer blends, where its brittle nature is improved by the incorporation of rubber fillers, particularly of the SBR or ABS type. It is also blended with polyphenylene oxide, with which it appears compatible,[4] forming the minor component at typically the 25:75 level of composition. In this way the processability and low cost of polystyrene draws on the excellent heat and dimensional stability of the other component.

Like polystyrene, polymethyl methacrylate is usually used because of its clarity, but on the other hand it is also commonly pigmented, and it is used to a much less extent in polymer blends. Acetal polymers are often compounded with carbon black or titanium dioxide as pigments and light stabilisers. To improve their bearing characteristics, they may be compounded with molybdenum disulphide or with polytetrafluoroethylene, typically in a composition of 25 parts of the latter to 75 parts of the acetal resin. Polytetrafluoroethylene itself, which suffers from excessive cold flow behaviour, high compressive creep, and wear, is filled with particulate fillers such as graphite, bronze, and molybdenum disulphide. Anticipating the next section, it is worth mentioning that dispersions of PTFE are sometimes mixed with glass and asbestos fibres.

Of the step-growth polymers, nylon plastics are frequently filled with mica, quartz, and finely divided graphite, whilst molybdenum disulphide is used to improve bearing properties under high loads and abrasive conditions. Silica is also used as a filler, loadings up to about 70 per cent being possible following appropriate surface treatment. Talc and glass beads also find use. Linear aromatic polyesters are available as moulding compounds containing talc, this also having an influence on the crystallisation behaviour of the polymer.[5] Another polymer which finds application because of its clarity as mentioned earlier, is polycarbonate, but pigmented grades and translucent silica-filled material is of importance.

The last composites to be mentioned in this section are those based on thermoplastic elastomers. Of these SBS and SIS block copolymers are not only compounded with other polymers such as polyethylene, polypropylene, and, particularly, polystyrene, but themselves may be compounded with calcium carbonate, silica, and clays to reduce cost and modify mechanical properties. Again, as already outlined, thermoplastic polyurethanes are also blended with other materials including other polymers.

Reinforced Thermoplastics

Although some particulate fillers such as talc and wollastonite do have a reinforcing action, the most significant effect in this direction is obtained through the use of fibrous fillers which take thermoplastic polymers well into the class of technical structural materials known as engineering thermoplastics[6] with mechanical properties which allow them to compete with and even excel metals. The use of reinforced thermoplastics long prefaced the first glass-reinforced polystyrene introduced in 1951, but the first really important fibre-reinforced thermoplastics, starting with glass-fibre-reinforced nylon, came along in the 1960s. Since this time the rate of growth, as a plastics material, has been one of the highest in the field of polymers, with glass-reinforced polypropylene being the growth leader, at least in the USA. From a production of 16 000 tonnes/annum in 1970, the figure for this system had risen to 95 000 tonnes/annum in 1979. World consumption of reinforced thermoplastics is now running at a few hundred thousand tonnes each year. One of the advantages of reinforced thermoplastics as compared with reinforced thermosetting polymers which is helping this growth is that they may be processed by high-speed injection moulding. Amongst the other polymers which are used as matrixes for fibre reinforcement, in addition to nylon which is particularly important in Europe, there are acetal resins, polystyrene and related polymers, polyethylene, polysulphone and other heat-resistant polymers such as polyphenylene oxide, polyphenylene sulphide, and the polyimides, polycarbonates, polyurethanes, polyesters, polyethers, and polyvinyl chloride. Most applications concern the bulk polymers, but structural foams are also of importance (see below). With reference to particular systems, glass-fibre-reinforced nylons such as nylon 66 and nylon 6, have increased toughness, abrasion resistance, tensile strength, and compressive strength. One of the problems in forming fibre-reinforced thermoplastics of this kind is the tendency of the fibres to orientate themselves during moulding. For this reason there has been a move towards the use of mixed particle–fibre systems, the particulate mineral filler counteracting the orientation. One of the effects of the addition of fibres is to raise the heat distortion temperature of polymers. For nylon 66 containing 40 per cent glass fibre the increase is such that at 280°C, the heat distortion temperature is above the melting point. Glass-reinforced polyethylene and polypropylene are much improved in stiffness, creep resistance, and heat distortion temperature, the latter polymer composite finding growing application in automobile fittings and in washing-machine parts. Polyesters, such as polyethylene terephthalate and polybutylene terephthalate, are also reinforced with glass; the latter, which like nylon may also have mineral filler incorporated, experiences an increase of heat distortion temperature of from about 50°C to 190°C in a typical formulation. Both polyester composites have better water absorption resistance compared with nylon polymer systems and have been used in water pipelines, although it should be mentioned that glass-fibre-reinforced nylon has found application in car radiator header tanks. Polycarbonate is reinforced with up to about 40 per cent glass fibre to provide tough, but still transparent shields for motor cycles, etc. Glass-fibre-reinforced polytetrafluoroethylene has been used in bearings and fibre-impregnated polymer in roofing, although this is not a conventional example of a reinforced thermoplastic. Illustrations of other applications are the use of glass-fibre-reinforced polyphenylene oxide as a heat-resistant solder bath material, ABS-glass reinforced polymer in sound damping, glass-fibre-reinforced polyvinyl fluoride in corrosion resistance, and reinforced polyphenylene oxide/polystyrene blends for piping. Glass-fibre-reinforced grades of a comparatively new thermoplastic polymer, polyether ether ketone, are on the market.

In addition to glass fibre, increasing interest is being shown in other fibres including carbon, asbestos, boron, aramid fibre, metal whiskers, etc. Carbon fibre is used in nylon, polysulphone, polyester, polyphenylene sulphide, polycarbonate, polypropylene, and fluorine polymers, typically in the composition range 20–40 per cent. A special attribute of carbon-fibre-reinforced polypropylene is that it is capable of being electroplated, of particular value in the automobile industry. Although it is expensive, as demand increases for its contribution to high specific strength and stiffness, and with its low expansion coefficient, the future seems to be promising, although carbon fibre for reinforcement of thermoplastics still faces competition from boron and aramid fibres. Since all three suffer from relatively high cost, perhaps the greatest immediate potential for these as reinforcing agents is in hybrid systems as will be discussed presently. Asbestos fibre gives a significant increase in heat distortion temperature and decrease in expansion coefficient. At about the 40 per cent level it has found use in heater casings when added to polypropylene. It has, however, an inherent deficiency as a fibre on impact behaviour, but this can be improved by appropriate treatment with a coupling agent. Aramid fibres have found application as reinforcement for polyurethanes and polyvinyl chloride in particular.

REINFORCED AND SYNTACTIC FOAMS

Strictly speaking, a foam is itself a polymer composite of discrete or continuous phases of a gas, typically air although in the short term it may be some other gas, within a polymer matrix, but in the context of this book, where the emphasis has been on rigid or semi-rigid fillers, foams have been excluded. Even so, we are still left with three types of foams which to some extent, do not fall within the spirit of the exclusion. The first of these is that type of expanded structure typically sandwiched between two rigid surfaces, which finds wide application as building panels. But since these are essentially macrocomposites, they too will be ignored for our purposes. However, they do have a significance in that, in their development it was not uncommon to use wires or synthetic fibres as internal supports for the panels, which leads naturally to a type of foam which will concern us, the reinforced foams. But before describing these, it is relevant to introduce the third type of expanded polymer composite, generally termed syntactic foams, which being particulate-filled will be outlined first.

Syntactic foams are composed of a dispersion of rigid hollow microspheres in a solid polymer. The spheres may be inorganic in nature, examples being glass balloons and fly ash, the latter a by-product of coal-fired power stations. Alternatively they may be organic, based on, for example, phenol and urea–formaldehyde resins, polyvinylidene chloride (Saran), and expandable polystyrene. The last are fabricated as small spheres containing a blowing agent which can expand at the softening temperature of polystyrene ($\sim 100°C$), which may be exceeded in the reaction endotherm of a particular matrix polymerisation. The microspheres used in syntactic foams are usually of a size corresponding to about 30–120 microns with wall thicknesses of 2–3 microns. The resultant foams are generally stronger than unfilled foams and have densities of the order of 300–600 kg/m^3. They can also withstand higher hydrostatic pressures than conventional foams. They find application, sometimes with added woven reinforcement, in epoxy, polyester, polyamide, polyurethane, polyvinylidene chloride, silicone, polyimide, polystyrene, and urea–formaldehyde resin matrixes as buoyancy aids in marine uses in ships, submarines, and boats, as buoys, cones, boards, rudders, etc. They are used in camping equipment, in electronic and microwave equipment because of low electrical loss properties, and in simulated light-weight wood applications.

Reinforced foams based on fibre reinforcement have been known for some years but do not seem to have maintained their early promise of a new generation of high-strength, very light weight structural materials. However, there is still considerable interest in them and in the recent partially porous reinforced polymers used in reaction injection moulding. The foams can be produced in different ways such as by adding expandable polystyrene beads to glass-fibre-filled epoxy resins, or more usually by injection of nitrogen gas into a reinforced polymer melt or through the use of conventional blowing agents in the same kind of way. Polystyrene foam, reinforced with glass fibre, can be prepared in the usual way as for ordinary polystyrene foam. Glass-fibre-reinforced polypropylene foam is used in brush head blocks, carbon-fibre-reinforced polyethylene foam finds application in the construction of fruit cases, glass-fibre-reinforced expanded ABS polymer is used in car seating, glass-fibre-reinforced expanded polystyrene is used in furniture, and glass-fibre-reinforced expanded nylon has been used in automobile panels and in self-sealing fuel tanks in military vehicles. A rather special application of a woven-glass-reinforced polyurethane foam was for insulation in Saturn V in which the foam helped to prevent debonding from the aluminium frame of the fuel tanks. The insulating property has led to application in trays. In the current interest in RRIM, short- and milled-fibre glass fibre has found wide use as has already been described. Before leaving this section on reinforced expanded polymers it might be of interest to draw attention to the way that fibres are distributed in the foam. Rather than lying at random with fibres perhaps protruding into the cells of the foam, they are found to be contained within the matrix with the directions of the fibre axis being in the plane of the cell walls[7] (Fig. 7.2). This will have particular influence on the mechanical properties.

FIG. 7.2. Alignment of reinforcing fibres in cell walls.

HYBRID COMPOSITE SYSTEMS

The use of mixed fibre and particulate composites has been recognised for many years and reference to this, in for example, unsaturated polyester and thermoplastic nylon and polyester moulding, has already been made. Some mention has also been made of mixed fibre systems, known as hybrid composites, and it is the intention of this section to say a little more about them in view of their great potential in the use of polymers in engineering applications. Growth of hybrid systems has been particularly rapid in the last decade or so, encouraged by the demand for high-performance engineering materials, especially where low weight is of great importance. Although individual classes of fibres can contribute some desirable property, the real interest in composites is in optimising the different contributions from different types of fibres, at the same time keeping an eye on reducing cost. An alternative approach, which does not as yet appear to have made a big impact, would of course attempt to have fibres of the same class available but for the fibres themselves to be prepared in different ways in order to offer a range of different mechanical properties. In this philosophy of blending fibre contributions, carbon fibre may well be chosen for the high strength, high stiffness, and low density which it offers despite its relatively high cost of production. On the other hand, glass fibre has a better fracture strain and stress behaviour providing improved toughness and impact strength, and it is, of course, much cheaper than carbon fibre. Even greater toughness can be derived from the use of carbon and aramid fibre hybrids. This has led to their use in golf club shafts within an epoxy matrix. Other matrixes used for hybrid fibre reinforcement include nylon, acetal resin, polyethylene, polysulphone polymers, and polystyrene, but it appears that epoxy resins find the widest use at the present time. In these, and in polyester and some vinyl resins, the hybrids give rise to materials which can not only compete with, but excel steel on a cost–performance basis, and still, at the same time, have the further advantage of low density. Other fibres which are used in hybrids include boron, different types of glass fibre, e.g. E and S glass, and silicon-based fibres, the latter in combination with carbon fibre being able to withstand temperatures of up to 1300°C. The fibres can be combined in different ways,[8] in addition to the many different ways that single fibres can be arranged in space as described in Chapter 1. These include intimate mixtures of the two fibres, either as short or continuous filaments, discrete layers of different fibres, layers of mixed fibres as for example in tapes, and in sandwich structures in which a core may be of one type of fibre and the surrounding layers of another. An important hybrid which is sold in tape form is aramid with glass or carbon fibre. Similarly, the fibres may be blended in woven form also. To a first approximation, for unidirectional hybrid fibre arrangements a form of the Law of Mixtures is suitable for describing the effect on modulus:

$$E_H = E_1 v_1 + E_2 v_2 + E_M v_M \tag{7.1}$$

where H and M refer to the hybrid composite and matrix, and 1 and 2 refer to the two fibres. The comparable relationship does not appear to be valid for tensile strength. Figure 7.3 shows a typical stress–strain

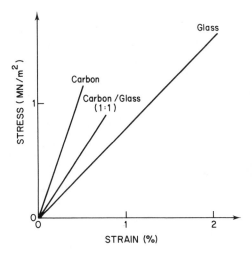

FIG. 7.3. Tensile behaviour of unidirectional carbon, glass, and hybrid (1:1) composites.[8]

relationship for such a system and compares it with the case for the separate fibres. It is believed that when the carbon fibres break in such a system of carbon and glass, the load is transferred to the glass fibres which allow further extension of the composite before total fracture occurs.[9] Dynamic fatigue properties of such hybrids are also improved.[10] In addition to the other properties contributed by carbon fibres they offer high thermal and electrical conductivity, good corrosion resistance, and reduced thermal expansion. One of the main factors which contributes to the current emphasis on composites is the reduced cost of hybridising with carbon whereas previously its high cost had restricted its use to special applications such as might be required by aerospace design applications, for example. Of course, the same principle will apply to boron and other expensive reinforcements. It is perhaps in this field that we shall see particular action in the next few years. Before outlining some of the applications of specific systems, mention should be made in the context of what was said above about predicting mechanical behaviour, that there is speculation as to the possible extent of synergistic behaviour arising from the use of hybrids, for example it might well be that aramid–carbon fibre epoxy laminates can have better impact properties than can be obtained with either fibre separately.[11]

Turning to specific applications, hybrid composites have been used in aerospace as in jet engine blades, in drive shafts, at least at the experimental level, in leaf springs in textile machinery and automobiles, the latter being of carbon and glass fibre. It is perhaps of interest to report that in the construction of drive shafts, epoxy resin impregnated unidirectional carbon and glass fibre is wrapped on to a mandrel, with particular attention being paid to control of resin content, winding angle, fibre tension, void content, winder speed, and cure temperature. Aramid–carbon fibre epoxy resin hybrids have been prepared for use at $8-16 \times 10^3$ rev/min. Nylon and carbon have been used in fishing rods and cross-country skis, whilst boron has also been used with carbon in sports equipment and aerospace parts, and with nylon for tennis rackets, as has also glass and carbon. In this type of application, as in helicopter blades, the core reinforcement may well be different from that of the outer layers, which indicates the further versatility which is available from geometric as well as material combinations. It is probable that current research in pultrusion fabrication methods as well as new resin systems may open up new and exciting outlets for hybrid systems.

POLYMER CEMENTS

Although not strictly relevant to the present text some mention should be made of polymer cements, since not only are they composites involving polymers but phase-inverted from what has gone before, but discussion of their properties may well take place at meetings primarily directed towards straightforward polymer composites and so the polymer scientist and technologist should be aware of their standing. In addition, by making some mention here, it is hoped that the reader will be led to appreciate that considerations on composites are not restricted to polymers but apply to materials as a whole, particularly since exciting developments are taking place in the fields of metals and ceramics, just as they are for polymers, and that the principles involved in their behaviour are basically the same irrespective of the material. However, it must be recognised that the properties of polymer cements are essentially modifications of those of inorganic cements, rather than of organic polymers.

Polymer cements can be prepared in three main ways. The first is to interdisperse a polymer emulsion within the cement, the second is to construct a cement using a water-soluble polymer, and the third way is to disperse a monomer in the cement and then polymerise the monomer actually in situ. The first method is the oldest way of preparing polymer cements and a variety of latexes have been used for this purpose, including natural rubber, neoprene, SBR, polyvinyl acetate, etc. Not all latexes are suitable and there have been problems with coagulation. One of the advantages of these systems is that because of the relative workability of the mixtures less water is needed than for conventional cement systems which tends to encourage higher strengths. In fact, it is not unusual to obtain compressive strengths 2–3 times that of hydraulic cement. Other properties which are enhanced are hardness and flexibility, the polymer layer being able to reduce stress build-up. It is of interest to remark here, in contradistinction to what was said above, that in a case such as this rather than having a situation of phase reversal, we have essentially an interpenetrating polymer network, one polymer being organic and the other, inorganic. The same continuity of the organic phase, even though it might be the minor component, was seen in polymer blends, where the minor component is much more fluid than the major component at the time of moulding. In addition to polymers being incorporated in the way just described, reinforcement through the use of polymeric fibres such as polypropylene and nylon fibres, is not unusual. In this they complement the traditional role of steel, asbestos, and alkali-free glass as well as of carbon fibre. Cements bound with water-soluble polymers are not of great industrial interest, but systems based upon water-soluble phenolic, urea, and melamine resins have created some interest as have water-soluble cellulose derivatives and polyvinyl alcohol. The other important class is that of the resin cements, in which a monomer or low molecular weight resin is allowed to diffuse into the cement before being polymerised. An example of this is a polymer-impregnated concrete prepared by drying a concrete and then allowing low viscosity monomer, such as methyl methacrylate, styrene or acrylonitrile to diffuse into the matrix and then polymerising this by free radical polymerisation. Mixing of the monomer at the original cement-mixing stage does not appear to produce useful results as far as properties are concerned, but mixing the inorganic aggregate with a reactive resin such as an epoxy resin, or, less usually, a polyester with styrene, etc., is used for the preparation of polymer cements. Like the first class of polymer cement, these materials find application as non-slip floors, grouting, decorative panels, repair cements for concrete roads, and for resurfacing. They have also been used for machine beds. Again in these, fibre reinforcement can be utilised and, indeed, the possibility of using urban wastes, especially of high glass content, has been examined for preparing such systems. Before leaving this section, although not immediately relevant, mention might be made of the use of bicomponent melded fabrics in stabilising movement of aggregate material in temporary road making, the effect here, however, not being a true composite one.

FUTURE DEVELOPMENTS IN POLYMER COMPOSITES

Predicting the future of polymer composites at this time of rapid social and economic change is not at all easy. Nevertheless, since the history of technology has shown that the elements of change, sufficient to influence at least a decade ahead, are very often already with us, an attempt will be made to identify these elements and assess their likely significance in the years ahead.

Possibly one of the most dramatic events in this context was the oil crisis of 1973–4 which caused social, political, and economic reverberations that continue even today. Not only did it alert the world at large to the fact that there was a limit to mineral oil sources but it also led to public recognition that oil was not the only raw material commodity in danger. It is predicted by UNIDO that by 1990, 15–20 per cent of petrochemical production will be in the hands of the 18 or so oil-producing nations, this corresponding to a synthetic polymer production of about 20 million tonnes. It is clear that the affairs of many of the present technologically advanced countries will be inextricably mixed with those of the developing countries. It might well mean also that the intensity of demand for plastics goods could move dramatically to those countries now producing this wealth.

Returning to the specific matter of polymer composites, one of the consequences of the oil crisis has been to boost their development for a number of reasons. The first was that although the price of petroleum naphtha feedstock, the main material source of the polymer industry, rose sharply, the effect on steel, aluminium, and glass, for example, was even more severe. Coupled with the fact that composite polymeric systems, especially those containing large quantities of inexpensive filler, cost much less to make, and yet, except for their rather limited thermal properties, often have better specific mechanical properties, it is no wonder that a few years ago the volume production of polymers surpassed that of steel. With a trend towards light-weight automobiles, this itself a tendency fostered by high fuel costs, the supremacy of plastics is again established, the modern motor-car containing about a hundred items of composite polymeric material. Currently, the average European car contains about 60–70 kg of polymer, equivalent to a weight saving of 200–300 kg. By the late 1980s it is estimated that the weight will be 100 kg leading to a weight saving of 300–400 kg.[12] In terms of running costs, this could be equivalent to a saving of over 7000 litres of fuel in the normal lifetime of a car. This directs attention to another consequence of the oil shortage: a number of countries are looking for alternative sources of fuel. Any change in the nature of the fuel might lead to a reappraisal of the nature of currently used polymers which may not have the same tolerance to the new fuels as they do for the old. It is not easy to see a direct impact on composites, and it might well be that a modification of existing polymers will suffice rather than having to discover alternatives. On the other hand an increase in demand for a given polymer could have an effect on price which could in turn influence its position in another field.

Although much of the impetus for the development of new high-performance composite polymers comes from the automobile and aerospace industries, much use is made of the products in other industries. This means that if there is an increased demand through increased prosperity in some countries, even though the *per capita* consumption may be well below that of some of the more technologically advanced countries, then the future for composites will be assured. Since increased demand could lead to a fall in the price of some of the fairly expensive high-performance polymers and fibres, perhaps assisted by newer and cheaper ways of manufacturing the latter, a growth spiral could easily be imagined. Other factors, at least in the short term, which could make a similar contribution include developments in hybrid systems and in improved coupling agents. Already materials regarded primarily as extenders a few years ago are now accepted as reinforcing agents or fillers which upgrade properties in some other way. The highest growth rate is expected to be with carbon fibre (~ 25 per cent), but even the commonplace fillers such as calcium carbonate, wollastonite, and glass spheres are anticipated to top a 10 per cent growth rate.

Although it is unlikely that we shall see many more new commodity polymers in the near future, present research into the possibility of alternative raw material sources might lead to starting materials which are more convenient for the production of one polymer rather than another. If this were to be the case, remembering that already coal is being used as a source material for the preparation of polyvinyl chloride in South Africa, then there might eventually be shifts in the relative output position of present-day large-scale production polymers. If one type of filler is closely identified with one class of polymer, then a corresponding shift in the relative position of the filler with respect to other fillers might well ensue. Again, although no new commodity polymers are on the horizon, each year sees the introduction of new and modified fillers, such as chemically and structurally modified glass fibres and high-modulus organic fibres, or particulate fillers coated with an improved coupling agent. Changes that will affect the use pattern of fillers are likely to take place in this area in the next few years.

Possibly a minor yet significant development in the coming years, which will influence the position of composites, could be a plastics waste recycling programme. Until ancillary activities associated with the collection and isolation of waste polymer become substantial commercial possibilities, the effect will be small. However, achievements in the compatibilising of polymer blends, perhaps leading to second-grade composites, or legislation by governments concerned about plastics waste, could stimulate growth in this area.

The last development region to be outlined here is that of advances in the processing of composites. The front runner in this at the present time is RRIM, but many other processes are being studied, and if production rates for thermosetting polymers, with their often better thermal properties, can be made competitive with conventional injection moulding, this might lead also to a change in emphasis. Research in this topic by specialised engineers working with all the control facilities offered by microprocessors, might very well change both the pattern of composite production and intrinsic properties and behaviour in the next decade.

REFERENCES

1. *Reinforced Plastics*, p. 110, 25 April 1981.
2. See WEATHERHEAD, R. G., *FRP Technology*, Applied Science Publishers, London (1980).
3. *Plastics and Rubber Weekly*, 2 September 1977.
4. WIGNALL, G. D., Oak Ridge National Lab., Tennessee, private communication.
5. JACKSON, J. B. and LONGMAN, G. W., *Polymer*, **10**, 873 (1969); IBBOTSON, C. and SHELDON, R. P., *Brit. Polymer J.*, **11** 146 (1979).
6. TITOV, W. V. and LANHAM, B. J., *Reinforced Thermoplastics*, Applied Science Publishers, London (1975).
7. WILSON, M. G., *SPE Journal*, **27**, 35 (1971).
8. SUMMERSCALES, J. and SHORT, D., *Composites*, **9**, 157 (1978).
9. LOVELL, D. R., *Reinforced Plastics*, 216 (1978).
10. PHILLIPS, L., *Composites*, **7**, 7 (1976).
11. DANIEL, I. M. and LIBER, T. in *Composite Materials: Testing and Design*, p. 330, 4th Conf., A.S.T.M. Philadelphia (1977).
12. *Plastics and Rubber Weekly*, 19 September 1981.

AUTHOR INDEX

Numbers in italic type indicate those pages on which references are given in full.

SUBJECT INDEX